Wired into Nature

THE HISTORY OF COMMUNICATION

Robert W. McChesney and John C. Nerone, editors

A list of books in the series appears at the end of this book.

Wired into Nature

The Telegraph and the North American Frontier

JAMES SCHWOCH

UNIVERSITY OF
ILLINOIS PRESS
Urbana, Chicago, and Springfield

Manufactured in the United States of America
1 2 3 4 5 C P 5 4 3 2 1
∞ This book is printed on acid-free paper.

Library of Congress Cataloging-in-Publication Data
Names: Schwoch, James, 1955– author.
Title: Wired into nature : the telegraph and the North American frontier / James Schwoch.
Other titles: Telegraph and the North American frontier
Description: Urbana : University of Illinois Press, [2018] | Series: The history of communication | Includes bibliographical references and index.
Identifiers: LCCN 2017037674| ISBN 9780252041778 (hardcover : alk. paper) | ISBN 9780252083402 (pbk. : alk. paper)
Subjects: LCSH: Telegraph—West (U.S.)—History. | Telegraph—Social aspects—West (U.S.)—History. | West (U.S.)—History.
Classification: LCC TK5123.6 .S39 2018 | DDC 384.10978—dc23
LC record available at https://lccn.loc.gov/2017037674

E-book ISBN 978-0-252-05045-9

For Travis

Contents

Acknowledgments

My deepest appreciation goes to all the librarians, archivists, curators, and staffs at the various institutions where I conducted research for this book. These libraries and institutions include the Northwestern University Libraries; the National Archives; the Smithsonian Institution; the Library of Congress; the Rush Rhees Library at the University of Rochester; the Massachusetts Historical Society; the Newberry Library; the Grove; the Alaska State Library, Archives, and Museum; and the Anchorage Museum. In addition, a number of institutions granted permissions to use images in their collections and assisted with providing those images. They include the front cover image, courtesy of the Taft Museum in Cincinnati, Ohio, as well as images provided by some of the aforementioned institutions and other institutions. Danny Nasset at the University of Illinois Press was invaluable as my editor. I also am thankful to the entire staff at the University of Illinois Press and for the comments on the manuscript by anonymous reviewers. Ellen Wartella, Chair of the Department of Communication Studies at Northwestern, provided department funding to support the production of this book, and her support and the support of my department colleagues enhanced the illustrations, layout, and overall book design, for which I am grateful. Mimi and Travis shared their enthusiasm, as always. My students gave me good advice about this project on many occasions, as did many friends and colleagues. My thanks to everyone.

Introduction

> A local weekly report has been furnished the Indian Journal at Eufaula, Indian Territory. The interest taken by the citizens of this place is gradually increasing, and the fear and awe with which the natives have looked on this office is slowly wearing off, and they are beginning to understand for what purpose this office was placed in their midst.
>
> —Sergeant George H. Crane, Signal Service weather observer, Fort Gibson, Indian Territory[1]

This book is a historical study of the growth and deployment of the telegraph in western North America, concentrating on the latter half of the nineteenth century. The impact of the telegraph upon the United States of America during the last half of the nineteenth century has not been lost on scholars who study media and communication.[2] Over the years, a number of distinguished books and articles have been written about the telegraph and the nineteenth-century American experience. For the most part, past scholarly work is geographically partial. The bulk of research looks from the Atlantic Ocean westward only to the Mississippi River, and from Canada southward only to the Gulf of Mexico. The standard histories of the American telegraph are stories of the East Coast and the Atlantic Seaboard, the growing Midwest, and service to urban areas. The accounts often emphasize creators, corporations, the Atlantic cable, and financiers. The narratives usually foreground news, politics, urbanization, and commerce. They are told in part as cycles of invention, innovation, manufacturing, and economies of scale. These are partial histories, narratives of the nineteenth-century telegraph and eastern North America. This book looks toward the West.

The story of the telegraph and the nineteenth-century American West includes establishing American electronic communication networks across challenging landscapes and ecosystems.[3] The construction and operation of military telegraph networks by the United States Army Signal Corps and then using the telegraph as a military asset to wage long-term low-scale continuous asymmetric warfare against Native Americans occurred throughout western North America. Congress instructed American private telegraph entities to serve national security needs. The nationwide expansion of electronic monitoring and surveillance and the genesis of the White House Situation Room emerge in this story. In addition, the expansion of commercial telegraph interests in western North America increased throughout the latter decades of the nineteenth century. All this and more regarding the telegraph began in the nineteenth century, not so much in eastern but in western North America. This happened west of the Mississippi River, in territories before they achieved statehood, on the Great Plains and trails, at forts and stockades, in the deserts and mountains, north to Alaska and the Arctic and south to the border with Mexico. Congress, the president, and the military continuously engaged these and more aspects of the telegraph and its relationship with the nineteenth-century North American frontier. Above and beyond all these events, the telegraph was bound up in a multitude of nineteenth-century efforts to know North America, particularly through natural history, geodesy, cartography, climatology, and the environment. These themes of landscapes, ecosystems, biota, and weather are the recurrent streams of this story.

Just as many have explored relationships between the telegraph and the nineteenth-century American experience, so have many explored relationships between the growth and development of twentieth-century American electronic communication networks and the growth and development of the United States as a global power.[4] Like the standard history of the nineteenth-century telegraph, this twentieth-century story of American electronic communication networks and the rise of American global power is often anchored on the Atlantic Ocean. The temporal trajectory then extends across the Pacific Ocean and eventually across the entire landmass of the planet. The growth proceeds across continents, on the beds of the oceans and the seas, and above the planet with satellites in outer space. Here too the story of the telegraph and western North America provides perspectives different from the centrality of the Atlantic as the geographic setting to anchor upon and witness the emergence of the United States as a global communication power spread across, below, and above the surface of Earth. With a focus on the nineteenth-century telegraph and the American West, the Arctic becomes a more central locale. Russian America, Canada, and Mexico loom larger. In particular, the story of the telegraph and the nineteenth-century North American frontier illuminates the transformation of American electronic communication networks and the transformation of American power and influence from continentalism to globalism. This telegraph tale does so in new ways not as fully narrated in previous scholarship.

None of this is to deny, denigrate, dismiss, or diminish the work of previous scholars working on media, communication, globalization, and American history. I am a huge admirer of that literature. Many of these works brilliantly explore the contours and interplay between capitalism, globalization, telegraphs, and cables.[5] Understanding capitalism and understanding globalization in the contexts of telegraphs and cables is illustrative and important. So too are approaches and viewpoints centered on the telegraph, cables, business history, and history of technology. The aim of this book is not to turn away from, disdain, or reject other scholarship about the telegraph, and neither is the aim of this book a call to abandon the insights and trajectories of those other works. Rather, the aim of this book is to offer new perspectives and viewpoints regarding the historical development of the American telegraph more centered on landscapes and ecosystems, more centered on questions of indigenous peoples and surveillance, and more centered on environment and climate. From time to time this includes a look at the telegraph in context with older notions of nineteenth-century knowledge such as natural history. Capitalism and globalization are visible in the story of the telegraph and western North America, and environment, climate, and surveillance are visible in the story of the telegraph and eastern North America. But more of the capitalism and globalization story is found in the telegraph and eastern North America, while more of the environment, climate, and surveillance story is found in the telegraph and western North America. In order to place environment, climate, and surveillance more closely to the center of the narrative, this book largely looks toward the telegraph and the American West. This approach also demonstrates that our contemporary world of the United States of America and global electronic communication with massive networks and massive data flows, cybersecurity and cyberwarfare, secrecy and surveillance, and complex relations between the federal government and telecommunications corporations has a deep and intimate history with the telegraph in both eastern and western North America.

Narrative Themes

This book advances four broad themes for considering relationships between the telegraph and the North American frontier in the latter half of the nineteenth century. These themes are (1) The High Ground, (2) The Signal Flow,

(3) The State Secret, and (4) The Secure Command. The book explores several aspects of these four themes in the contexts of Native American conflict, expanding telegraph networks across the Trans-Mississippi West, military and commercial telegraph networks, and the significance of codes and ciphers. Arctic exploration and the development of territorial Alaska and the role of the telegraph in plotting longitude and time zone coordination across the continent are included. The growing realization by a range of officials in the federal government that telegraph networks and other electronic communication systems (such as the telephone) needed to be constructed and maintained as secure systems and means of communication took root. Always, recurring, and continuous, the telegraph-enabled pursuit of scientific, environmental, and climatological knowledge about the continent proceeded apace. The analysis concludes with the 1898 Spanish-American War and the transformation of the nation from a continental to a global power. The narrative demonstrates that these four themes of the telegraph and nineteenth-century western North America provide an explanatory context for this continental-to-global transformation at the dawn of the twentieth century. These factors and this narrative of the telegraph and the nineteenth-century American West help account for a long historical arc of various contemporary issues having to do with electronic surveillance, privacy, data gathering, and relations between the federal government and telecommunications, media, and information technology corporations. They also help account for a long history of seeking and deploying continental (and eventually global) knowledge about the environment, climatology, and weather.

The High Ground recalls the axiom that securing and controlling the elevated terrain of a particular area is a key element for successful military control and security of that area. Outer space is now often referred to as "the high ground" in terms of contemporary global security. Securing the high ground for the telegraph in nineteenth-century western North America involved the successful mastery of difficult terrains, such as mountains and deserts. The high ground also invokes the landscapes and ecosystems of western North America. These landscapes and ecosystems posed a distinct set of challenges for the construction and establishment of telegraph networks. The relative lack of trees (for telegraph poles) on the Great Plains and in the Southwest, the scarcity of timber, the absence of local manufacture, and the relative lack of both a navigable riverine system and a system of post roads, all before the railroad spanned the continent, made the introduction of telegraph networks in western North America arduous. A vast range of weather conditions, challenges in transporting equipment, and finding suitable forage for draft animals were routine problems. Securing the high ground also meant the telegraph became a key sociotechnical infrastructure in the transformation of knowledge about the landscapes, ecosystems, and indigenous civilizations and cultures of the North American frontier. This included telegraph-assisted geodesy (spatial measurements of the surface of the planet) to fix longitude and coordinate time zones. The high ground invokes environmental and climatological factors, nation-state control, indigenous populations, flora and fauna, and the landscapes of western North America.

The Signal Flow is arguably the fundamental element of any electronic communication network. Without the flow of signal, there is no possibility of effectively communicating throughout the network. For the telegraph, this included reach, reliability, durability, and speed. Maintaining a constant and reliable signal flow for the telegraph throughout western North America was beset by such challenges as extreme weather, resistance, conflict with indigenous peoples as well as criminals and outlaws, and a near-constant rebuilding and rerouting of telegraph lines. Another important issue regarding signal flow was a nineteenth-century version of net neutrality: should the signal flow of one party on a telegraph route or network take precedence over the signal flow of another party? For government uses of commercial telegraph networks, could commercial corporations be required to provide the federal government with rates and a fee structure at a lower cost than other customers, and could the federal government have both priority access and the power to reconfigure circuits and signal flows on commercial networks to meet national security needs? Congressional hearings in the 1870s indicated that the answers to these questions were "yes." Beyond these questions concerning privacy and national security, the signal flow was also an ever-present condition regarding the continent-wide data collection of weather and climatological information (among other collections, such as coordinating natural history specimen collections for the Smithsonian Institution), and the dissemination of that information to Washington for storage and analysis. Reliable signal flow could and did protect property, assist agriculture with prompt weather warnings, and even save lives, particularly along seacoasts and on the Great Lakes. The signal flow encompassed nineteenth-century versions of contemporary issues such as global reach, net neutrality, secrecy and security, climatological and environmental awareness, and the ability of the military-government-intelligence system to dictate to commercial telegraph services—discreetly, of course—the national security needs of signal flow.

The State Secret recalls a range of issues. Codes and ciphers are one prominent theme. Military communication necessitated codes and ciphers that were not only secure within military networks but also secure over commercial networks. In 1872 the Signal Corps introduced a new cipher (to save costs) that caused consternation for Western Union. The new cipher was not a secret kept from Western Union, for the cipher was provided to commercial telegraph companies. Despite requiring fewer words, letters, and numbers than the previous cipher, however, the new cipher did not reduce transmission time but actually increased transmission time, thus raising the expenses of Western Union in carrying military messages. The use of the new cipher necessitated repeating many messages to reconcile errors, because the cipher relied on similar written code words that were difficult to aurally distinguish from each other in audible Morse code. That made it challenging to telegraphically relay messages with accuracy. Maintaining the confidentiality or security of a telegraph message was often compromised by shoddy practices in the storage or disposal of written messages, the common tendency to dictate messages aloud in a telegraph station in the presence of others, or lively gossip. The telegraph in western North America was a topic of local conversation, and conversations about individual telegrams often ballooned into speculation, bragging, rumors, and hyperbolic newspaper accounts about what was, or what was not, sent over the telegraph. The state secret was also embedded in discussions between the commercial services, the military, and Congress regarding the hypothetical possibility of future hostilities with other nations, and the subsequent conditions that war might warrant for the telegraph. While most parties saw this conversation as theoretical, a few saw this conversation as pragmatic: perhaps even an ongoing condition of containment and conflict regarding Native Americans, and of border security with Mexico, rather than a future possibility involving foreign powers. Certain views regarding the power of the federal government over electronic communication networks during times of conflict, surveillance, or an ongoing war on terrorism have historical roots in the rise of the Signal Corps and their doctrine and approach to ongoing and continuous conflicts between the army and Native Americans in the last four decades of the nineteenth century.

The Secure Command demonstrates that the importance of secure electronic communication networks for the president of the United States and the origins of what is now known as the White House Situation Room are not found in the 1962 Cuban Missile Crisis, but instead in the latter decades of the nineteenth century, spanning presidencies of Rutherford Hayes to William McKinley. The secure command of the telegraph was a highly contested issue during the Civil War. President Abraham Lincoln did not have a telegraph office in the White House and instead conducted telegraph communication at the War Department telegraph office in Washington, routinely monitored by War Department officials. The 1870s, 1880s, and 1890s all saw significant electronic communication network infrastructure upgrades in the White House and Capitol and slowly increasing autonomy over telegraphic communication and electronic communication networks for the president. The secure command was also found in the field. Securing the high ground and maintaining the signal flow ensured successful conditions for the state secret. This included battlefield tactics, and included the successful deployment of what might retrospectively be called new media and communication technologies (such as mirror-signaling with the heliograph, or in another example, the field telephone) when those new technologies proved useful and resilient. The secure command is discernible in both the network infrastructure associated with the highest levels of American government (such as the White House and Capitol), as well as on the battlefield, and new technologies can augment the secure command. The secure command enhanced the control of the high ground, signal flow, network security, and reach of secrets over various electronic communication networks. Yet nothing about the secure command was immediately self-evident. It was a contingent process, navigated in fits and starts over a span of forty-odd years. The secure command of electronic communication networks, culminating in the deepest centralization and highest authority of that command in the office of the president, was not immediately present but constructed over decades.

The four themes intertwine throughout the book in a narrative structure that incorporates landscapes, ecosystems, biota, and weather. This approach, evident for example in the chapter titles, undergirds a central factor in the growth and deployment of the telegraph in nineteenth-century western North America: continent-wide weather forecasting and data collection by the U.S. government. This began systematically with the Signal Service (a branch of the Army Signal Corps) in 1870, later transferred circa 1891 to the newly formed National Weather Service in the Department of Agriculture (NWS, now NOAA, the National Oceanic and Atmospheric Administration). This daily collection, dissemination, storage, and analysis of latter nineteenth-century continental weather data is now redeployed by contemporary scientists as one of the master historical databases for the long-term global analysis of climatology, oceanography, environmental variation, and climate change. Along with ice-core samples, tree rings, measurements of disappearing glaciers, satellite imagery, sedimentary layers, animal tracking, and similar means of data collection and analysis, the daily weather measurements of the Signal

Service (and later NWS and NOAA) dating back to November 1870 continue to contribute vital knowledge about the changing conditions of the planet. Chief Signal Officer Albert Myer proved prescient in his 1879 annual report when he forecast "It is upon the data now accumulating upon the files of this office . . . that the future studies of the climatology and meteorology of this continent will be based."[6]

Issues spurring the wiring of the American West included exploration, natural history, gold, Native American containment, frontier settlement, and visions of Manifest Destiny. No factor, however, was more central than the accumulation of environmental and climatological knowledge about North America and its weather. The investment in daily, monthly, and long-term weather data collection and analysis was mammoth. The investment influenced infrastructure issues such as new telegraph stations, new lines, and new networks interconnecting the territories and regions of western North America with each other, with the eastern United States, with Canada, Alaska, and the Arctic, with the Caribbean, via undersea cables with Europe, and most especially with federal offices and edifices in the nation's capital. Government weather forecasting, justified as a national security measure as well as a benefit to agriculture and commerce, proved crucial to the growth of the telegraph across the North American frontier and, by extension, the entire continent.

The Telegraph and Continental Knowledge

Finally, the theme or thread weaving together the entire period of study herein is the continual pursuit of continental knowledge and the growing mastery of knowing and acting upon the nature of North America. It is above all else manifest in Native American relations, natural history, climatology, and the environment. The growth and uses of the telegraph in nineteenth-century western North America are interwoven with the pursuit, collection, analysis, and application of continental knowledge toward mastery of North America. This mastery entailed governance, commerce, urbanization, economics, finance, and technology, as seen throughout eastern North America. Yet it also entailed Native American containment, security, frontier settlement, cartography and geodesy, and most of all the nature of the continent as it could be known through data collection and analysis regarding natural history, the environment, and the weather. Collected in vast quantities across the entire continent, the dispersed knowledge from across North America was delivered by telegraph (and other means) to Washington for analysis, storage, retrieval, and application. In this sense, this book also is about the emergence and growth of Washington as a data depository for the nature of the continent.

As it turned out, knowing North America in these ways proved advantageous for continental mastery, and it also proved advantageous for the global presence and growth of the United States of America in the late nineteenth century. Cleveland Abbe, a leading American meteorologist of the era, advised his colleagues in Europe that to know the present weather of North America is also to have advantageous insights into the probable future weather of Europe—but not the other way around. Abbe also told his European colleagues that if the natural world were different, perhaps they could generate more of their own knowledge. All that would take is their own weather stations on a string of islands properly situated on a north-south axis in the North Atlantic, three hundred or so miles off the European shore. Of course, no such string of North Atlantic islands existed. Knowing the nature of North America had a global impact. What now enhanced that impact was the telegraph, along with undersea cables.

It was one thing to be in Europe in the early nineteenth century and know North American weather trends and patterns as information was relayed, weeks after the fact, by transatlantic maritime traffic. It was quite another thing to be in Europe in the latter nineteenth century and know North American weather trends and patterns when daily weather forecasts were immediately relayed throughout the day and night, day in and day out, from Washington to Europe by cable and telegraph. Transatlantic weather information between Europe and North America had been gleaned for three-hundred-odd years from accounts of transatlantic sailing ships and whatever information they might bring back to port after weeks or months at sea. The cycle of the seasons was known, the likelihood of fog, winds, ice, or clear weather in a given locale during a particular season was general lore, but little else was known to any degree of reliability. By the last decades of the nineteenth century, North American weather data providing "a fairly accurate knowledge of the weather conditions prevailing over the great storm region of the North Atlantic" was cabled daily to Paris and London and further disseminated on the European continent. "Synchronous observations, storm gales, derelict wrecks, and dangerous ice noted for the previous five days by steamers" were included in the daily cable report.[7] That was the difference between knowing the likely weather in North America, in Europe, and on the North Atlantic Ocean for the next few days, compared with learning weeks after the fact why the weather in a particular locale was the way it was.[8]

Daily transatlantic data traffic about North American weather represented a new form of value sent from the United States to Europe. Goods, manufactures, the potential availability of land and employment to new arrivals, gold, and agricultural production such as cotton and food grains from North America all held value for Europe. Daily weather forecasts, transmitted by telegraph and cable, represented something new and different. This was measurement, data, logs and records, applied science, and the first major global manifestation of American "soft power" conveyed by electronic information networks. It was a form of knowledge transfer that transcended the previous three-hundred-odd years of data collection and knowledge dissemination by North American naturalists exploring landscapes and ecosystems, collecting specimens, making illustrations and maps, publishing their findings, and corresponding with their European sponsors and counterparts.

The European savants of transatlantic weather and climate were one of many communities with an abiding interest in North American weather and the work of Cleveland Abbe and the Signal Service. Newspapers, railroads, boards of trade, fisheries, sailors and river men, and farmers paid attention. In the 1880s, visitors to Signal Service weather observation stations west of the Mississippi included land seekers in Dodge City, butchers, house painters, and cotton merchants in Galveston, poulterers in Keokuk, and cattle drovers in Winnemucca. Some people visited a Signal Service weather observation station to calibrate their own barometers with the station barometer.[9] Even the iceman came to visit. For fisheries and maritime commerce, the "humbler industries of both the oysterman and the fisherman" found guidance from Signal Service weather reports "in the taking and in the preserving of the food gathered from the sea." Cattlemen warned of storms could now "save their herds from stampede." Farmers who were the "growers of vegetables and small fruits on which our great cities depend for supply" were now forewarned of early frosts.[10] By the late 1880s, a sufficient number of Signal Service weather observation stations had been established along the Pacific Coast to allow for timely warnings "in the raisin-drying districts of California at the period of earliest rains, which, without such warnings, are liable to do a vast amount of damage to the half-cured grapes."[11] Some saw a Signal Service station as a sign of increasing civility in their locale. The Signal Service operator in Pembina, Dakota Territory, believed the arrival and growing reputation of the Pembina weather observing station assisted the transition from the frontier to civilization. This was part of a welcome local trend toward amity: "the rowdy element is in a great measure disappearing before the influx of a better class of citizens."[12]

The Signal Service was not without its critics despite its efforts to boost its own social and economic worth. When Chief Signal Officer General William Babcock Hazen died in January 1887, the *St. Louis Republican* opined it may have been a "misfortune" for an officer of Hazen's abilities to be "drawn into a bureau in Washington" and to be held accountable by the general public for weather developments "that made him a target for newspaper shots every time the wind took a queer freak."[13] Yet many saw promise in the work that was yet to come, work indicating a future of growth. Abbe believed that the telegraph and the nationwide dissemination of daily news about the weather portended a future of ever-increasing interest, support, and expectations from the American people. "I think the more we attempt to respond, and the better we do our work, the more they want of us. . . . I anticipate the time when *any* and *every one* shall have the right to pay ten cents and demand his special weather telegram precisely as he pays his two cents and demands to have his letter carried by mail."[14] Many of today's social media users would agree.

Washington, the District of Columbia, and Telegraph Security

In the 1870s, architect of the Capitol Edward Clark, Signal Corps founder Albert Myer, Washington telegraph and telephone entrepreneur George Maynard, and head of the White House Telegraph Room Benjamin Montgomery were among many individuals who pursued telegraph and telephone security for the White House and the federal government in the latter half of the nineteenth century. The quest for secure electronic communication facilities for Congress, the president, the federal facilities of the District of Columbia, and the nation as a whole was not self-evident from the creation of the telegraph. Rather, the quest for secure electronic communication facilities within Washington, across North America, and extending to the furthest reaches of the planet and beyond was and continues to be an ongoing quest. Edward Clark, the first individual to officially hold the title of Architect of the Capitol, took significant steps throughout his tenure to ensure telegraphy (and later, telephony) would be secure electronic communication networks for the Capitol and for other District federal buildings. This imperative continued with subsequent architects of the Capitol.

The Capitol was not the only federal building to experience telegraphic and telephonic transformations in the last decades of the nineteenth century. The White House did not have a direct telegraph line until 1866 and did not have a full-time staff telegrapher until 1877. President Abraham Lincoln, who during his tenure in office sent more than a thousand telegrams, routinely walked over to the telegraph office of the War Department to send, receive, and read messages.[15] In 1877, at the request of President Rutherford Hayes, the Signal Corps assigned Benjamin Montgomery to the White House as a telegraph officer. Over the next twenty-eight years, Montgomery oversaw the expansion of the White House Telegraph Room into what one reporter writing in 1902 for *Telegraph Age* called "perhaps the most complete bureau of confidential communication in the world."[16] Having arrived just before the 1877 Railroad Strikes, during which Hayes used the telegraph in nationwide coordination with the army and the Signal Service to monitor and contain labor strikes and political unrest, Montgomery proved indispensable to the Executive Mansion. Hayes himself had been ushered into the presidency in part because of the telegraph and the astute political management of telegrams and telegraph lines during the presidential election of 1876. During the 1870s, Albert Myer transformed the Signal Service weather offices in Washington into a nationwide weather forecasting center and, in the process, turned that office into one of the biggest recipients and senders of telegrams in North America. Weather data flowed into that office at rates of more than fifteen hundred reports each day. Reports, forecasts, updates, and predictions flowed from the office each day to newspapers, railroads, local communities, government agencies, and other individuals and institutions across North America and around the world. By the 1880s, the Signal Service network generated hundreds of thousands of telegrams each year. George Maynard was crucial during the 1870s and 1880s in configuring the national telegraph routes and circuits for the Signal Service, and also crucial in bringing the new technology of the telephone to the White House, the Capitol, the federal government, and the District of Columbia.

The roles of Edward Clark, Albert Myer, George Maynard, and Benjamin Montgomery are important in this story, but their contributions represent only a few of a wide range of important interactions between Congress, the presidency, the military, and the telegraph in the nineteenth century. Congress authorized Samuel Morse to install his experimental telegraph system in the Capitol, leading to the well-known 1844 demonstrations by Morse. Through the Pacific Telegraph Act of 1860, Congress approved funding for the Transcontinental Telegraph, completed late in 1861 as the first line to traverse the continent. The Transcontinental Telegraph soon played a role in expanding the Union. The constitution of the state of Nevada was sent entirely by telegraph to President Lincoln in October 1864, thus completing the transition from territory to statehood. This telegraphic consummation arrived just in time for Nevada to cast its three Electoral College votes in the 1864 presidential election for Lincoln. The Nevada Constitution, a 16,543-word telegram, is the longest telegram preserved in the National Archives.[17] The Transcontinental Telegraph also meant that the 1864 election was the first presidential election in the United States to have what might be called an early version of nationwide real-time election returns.[18] Election trends, reports, updates, and results from across the Union were telegraphed immediately to the War Department in Washington. The president spent much of 1864 election night in the War Department telegraph office.

Congress regulated the telegraph, for example with the National Telegraph Act of 1866 and subsequent acts, many of which enabled the building of telegraph lines and networks along post roads and railroad rights-of-way. Representatives and senators learned the many possibilities the telegraph presented for politics, campaigns, and elections. Debates, hearings, committee meetings, resolutions, petitions, editorials, and speeches about whether the federal government should, or should not, nationalize the telegraph were a congressional staple during this era. To return to the telegraph infrastructure and communication security of late-nineteenth-century Washington, the Capitol and the White House routinely communicated by telegraph, including a direct private connection between the president and vice-president when the vice-president presided over the Senate.

Yet there is more to this nineteenth-century story of the telegraph, the Capitol, the White House, Congress, and the president than the interesting examples of physical infrastructure and communication security, electoral politics, regulation, and the expansion of the telegraph across the continent. Along with the military and the presidency, Congress throughout this era was a crucial factor in establishing the telegraph as essential to national security, regardless of public or private ownership and operation of telegraph facilities and networks. In 1872, Chairman of the House Appropriations Committee James A. Garfield presided over a hearing on the working relationships between the commercial telegraph companies and the Signal Service.[19] The committee found that the Signal Service, and by extension the federal government, had the inherent right to require commercial telegraph companies to send military and government telegrams over whatever circuits and routes the Signal Service or any branch of government dictated. Whether a Signal Service telegram took a direct or so-called zigzag route through the Western Union telegraph network (or other commercial telegraph networks) did not matter, even though a zigzag route often included a Signal Service directive for additional "drops" or copies of the telegram as it zigzagged its way to Washington. The drops were to be provided free of charge, at no compensation beyond the cost of a single message.

The route directed by the Signal Service, whether direct or zigzag, no matter how many additional circuits or relays that route required, was not to trigger additional compensation.[20] On this and many other occasions, Congress, the military branches, and the White House encouraged the telegraph corporations to pursue friendly and productive relations with the federal government in pursuit of common interests. National security, economic development, rapid military communication, low-cost government expenditures for commercial telegraphy, continental knowledge, and the business success of the American telegraph industry could and should be conjoined goals.

Congressional views, opinions, and decisions about the telegraph and national security in the nineteenth century emerged through the lived everyday experiences of many in the federal government as well as through hearings and deliberations about the working relationships between military and commercial telegraphy. These views, opinions, and decisions also emerged from a wide array of additional perspectives: transportation systems, urbanization and industrialization, nationwide communication networks, and fiscal and economic growth. Since the end of the Civil War, this growing awareness was evident about the telegraph and eastern North America. A growing awareness also was evident west of the Mississippi, but evident through issues having less to do with finance, politics, economic growth, and urban progress. In western North America, the value of the telegraph was in gathering environmental and climatological data, containing Native Americans, and government influence over commercial telegraph networks. The telegraph helped determine longitude and plot geodesy, aiding the extension and completion of continentalism and the transformation to globalization. A nascent awareness emerged that the telegraph could be useful for nationwide monitoring and surveillance of labor dissent and political unrest. The growth and deployment of the telegraph in the nineteenth-century North American frontier and across the whole of the nation was bound up with the growth and deployment of surveillance across all North America and the whole of the nation. Weather forecasts, specimen collections, cartography and geodesy, labor dissent, criminals crossing the border between the United States and Mexico, and Native American surveillance had some commonality. As James Fleming observed, "to separate meteorological from political and military data gathering" enabled by the telegraph "is to introduce an artificial distinction."[21]

Determining longitude meant establishing continental time zone coordination, which further meant, for example, an army stockade reporting on Native American activities would no longer telegraph "Indians seen riding east at sunrise" but now report "Indians seen riding east at 6:42am." In 1884, the Signal Service reported its "total length of 2,803 miles of frontier line . . . is one of the most effective safe-guards against Indian raids and war-like movements, since it enables the scattered military forces of the United States to obtain timely notice of such movements, and concentrate quickly at any point of attack." The Signal Service telegraph network in the Trans-Mississippi West had become "an engine of civilization" that enabled "the Government to throw an aegis of protection over the rapidly expanding wave of western emigration."[22] Meteorology and commerce were important to the Signal Corps. Native American containment and conflict, however, was equally if not more important and was always at the core of strategic interests and national security for the military telegraph system of the Signal Corps. An 1899 report emphasized that the military telegraph was "built for the better protection of immigration and frontier settlements from depredations, and especially for maintaining quick and sure means of communication for military purposes. No use will be permitted of these lines that would defeat the object of their construction."[23] Plotting latitude and longitude enhanced the ability of the federal government to contain Native Americans within the perimeters of geodetically plotted boundary stakes.[24] Congressional funding for cartography, topography, geodesy, and longitude played a role in profoundly changing ways of knowledge and lived experiences about the ecosystems and landscapes inhabited by Native Americans. The telegraph, both directly and indirectly, was part of the transformation of indigenous peoples wrought by their migration from the Great Plains to Standing Rock and their subsequent forced settlement. The telegraph in western North America became as interwoven with a wide range of issues and concepts invoking Native American containment, conflict, and national security as with landscapes, ecosystems, and nature. This is what it meant to be wired into nature: a particular construction of nature and power, and a peculiar institution of the telegraph and the nineteenth-century North American frontier. Knowledge, growth, pursuit, and practice of surveillance seemingly appeared natural. Nature provided an incredible bounty, including, as it turned out, core concepts of national security. Watching people and watching nature shared common ground.

Telegraph Lines and Railroad Routes

The growth and development of the railroad in western North America influenced and assisted the growth and development of the telegraph in western

North America. The reverse is also true, and the ancillary support provided by the telegraph to the railroad included coordination in the transportation of goods and passengers, the coordination of schedules for trains on tracks, greater safety and efficiency of railroad operations, and an additional revenue source for the many railroads that established telegraph lines over their rights-of-way. Repair and recovery operations after a derailment or construction of a rail line sometimes saw the installation of a temporary telegraph station at the derailment or construction site.[25] The railroad provided ancillary support to the telegraph with transportation of the physical components of telegraphy and of scientific instruments for Signal Service weather observation by the Railway Mail Service. Signal Service weather reports and the posting of Signal Service weather forecasts at railroad stations and depots, and the new telegraph lines owned and operated by the railroads, all benefited telegraph network growth and capabilities. A railroad right-of-way was an excellent route for a telegraph line, and railroad stations often were telegraph offices. In the 1870s and later, railroads proved much more effective for hauling telegraph construction materials across western North America than the wagons and draft animals used in the 1850s and 1860s.[26] In the late 1870s and early 1880s, railroads in western North America that also built a telegraph line along the right-of-way increasingly brought railroad time to

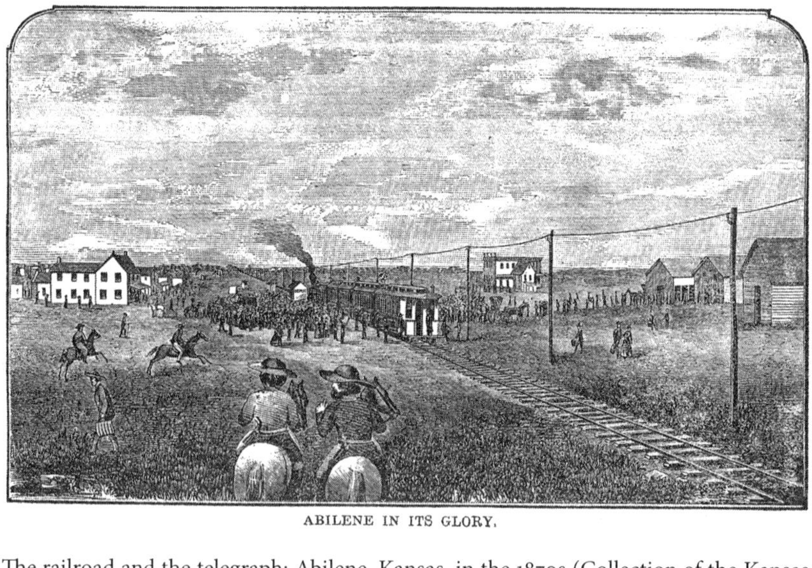

ABILENE IN ITS GLORY.

The railroad and the telegraph: Abilene, Kansas, in the 1870s (Collection of the Kansas Historical Society, item number 209714)

nearby towns and also to nearby telegraph networks and systems. This result was a valued commodity for commerce, science, and military actions.[27] The relationships between the railroad and the telegraph in western North America were systemic and synergistic.

The synergy of the railroad and telegraph was prevalent in nineteenth-century western North America. There was one important difference between the railroad and the telegraph: connectivity with the rest of the continent. Although both the railroad and the telegraph of course increasingly interconnected western North America with many cities and areas east of the Mississippi River, the interconnectivity point for the railroad and western North America that ranked above all other places and locales was Chicago. For the telegraph and western North America, the interconnectivity that ranked above all other places and locales was Washington. Chicago was the railroad nexus of goods, trade, and commerce flowing into and out of western North America. Washington was the telegraph nexus of data, knowledge, and security flowing into and out of western North America. Chicago truly was, as William Cronon wrote, Nature's Metropolis.[28] For the telegraph and western North America, Chicago was not Nature's Metropolis. There was a different metropolis for nature and the telegraph: Washington. The difference between goods and data generated a Nature's Metropolis for each city. Chicago was the railroad nexus, and Washington was the telegraph nexus.

Of course, goods and data flowed in and out of other metropoles. When eastern and western North America are conjoined as one national unit, New York City was the most important telegraph metropole for the nation as a whole. Finance, news, business activities, and a huge range of telegraphic data ceaselessly flowed into and out of New York City in the latter nineteenth century. New York City was the most important node conveying North American telegrams to the cables beneath the Atlantic Ocean, the cables that connected North America and Europe. New York City was extraordinarily important for finance, journalism, publishing, press agencies, and wire services. Yet for western North America, Washington was the telegraphic data depository transcending all other American cities. Information about weather reports, Native American containment, territories and their transitions to statehood, Smithsonian collections, cartography, geodesy, and military actions on the frontier all flowed by telegraph in and out of the nation's capital. The common currency for both Chicago and Washington was the nature of western North America. The currency of American nature had two technological manifestations: a technology for networking goods, and another technology for networking data. Chicago and the railroad represented one manifestation, and Washington and the telegraph represented a different manifestation. Both manifestations of the natural currency generated endless treasures from the same incredible wellspring.

Contemporary Trajectories

In exploring this history of the nineteenth-century North American telegraph, interesting and perhaps unexpected trajectories emerge from the experiences of western North America. These trajectories suggest that the experiences of the nineteenth-century telegraph in the Trans-Mississippi West offer a distinct history that helps account for our contemporary electronic media society with its attributes of surveillance, secrecy, data collection, and a complex relationship between the federal government and information technology corporations. Secrecy, surveillance, security, information networks, monitoring, data collecting and data privacy, and a host of related issues currently command significant social attention in the United States. In particular, the questions revolve around communication security and governance. These issues are exemplified by events such as the tragedy of September 11, the Patriot Act, the Snowden Affair, Stuxnet, Hillary's e-mail, Trump's tweets, WikiLeaks, FBI vs. Apple, a plethora of cloak-and-dagger dramas warning of compromised cybersecurity, reports of colossal credit-card customer data theft, and the strange tale of Sony and the North Koreans. Many Americans find these stories and events a routine part of their daily news feed, a somewhat confusing question regarding the balance of national security with individual rights and liberties, and a constant condition of their data-driven lives. Banking, work, leisure, health care, travel, politics, popular culture, and maintaining social relations are just a few examples of the data-driven life.

Some are aware of or otherwise recall censorship practices during World War II. A few know about communication security and Cold War issues. Historians and scholars have produced work tracing these questions back to, and even before, World War I. The telephone companies and the new wireless radio companies of the early 1900s pursued national security with the federal government. As it turns out, questions of electronic communication security and the nation as a whole go all the way back to the origins of the telegraph. Perhaps it is surprising that these questions are particularly evident when one looks at the telegraph and the North American frontier in the nineteenth century. Yet it is not so surprising after all. It appears that expansion—the nineteenth-century territorial growth and rising influence of the United States first on the North American continent, and then the twentieth-century transformation of that continental growth and influence into a worldwide American presence seen through economics and trade, popular culture, military activities, and additional factors suggest this is a continuous history. Communication security and electronic communication networks have something to do with the United States of America as a continental, and then later a world, power.

Terminology

This passage briefly discusses a number of terms used in this book.[29] The term *synoptic* appears in the middle and latter sections of the narrative. The usage here connotes *synoptic meteorology*: the assessment and forecasting of weather over a given area based upon synchronous sample measurements and data collection within that area. Suppose, for example, weather reports taken one day at 12:00 noon from Boston, Philadelphia, and Washington all indicate rain. Even though New York City was not sampled, it is reasonable to conclude there is probably rain occurring at the same time in New York City, as New York City is within a linear area (a storm front) bounded by Boston, Philadelphia, and Washington. The "probability" is key, however. One cannot be certain it is also raining at 12:00 noon in New York City without a local measurement that is simultaneous with the measurements taken in Boston, Philadelphia, and Washington. The more simultaneous sample points within a given area, the greater the probability of predicting rain within the entire area bounded by the furthest points of observation (sampling). Synoptic meteorology uses synchronous samples or measurements over a given area to infer the weather probabilities for the entire area, including points within the area not sampled. This also has implications for the idea of synoptic surveillance: synchronous measurements sampling a given area discerning the probabilities of labor unrest, criminal actions, or Native American activities within the entire area bounded by the points of simultaneous measurement.

Geodesy appears herein as a term on several occasions. Originally referring to the measurement and surveying of land, geodesy still carries these connotations. Specifically, geodesy is a branch, or application, of science and mathematics measuring the shape of the earth, including certain areas on the earth. Geodesy—or a similar term, geodetics—is based on astronomical calculations of time. It is related to the determination of longitude. For the telegraph, the ability to accurately coordinate local time with the time standard set at an astronomical observatory—essentially, the Greenwich

Observatory—allowed for the precise determination of longitude at a given point. In particular, this precise determination of longitude and latitude over a given area or space (that is, spatial knowledge of geodesy) was crucial for the precise delineation of the boundaries of Native American reservations by longitudinal and latitudinal coordinates. This allowed for the determination of reservation boundaries by geodesy, rather than natural landmarks such as mountains, rivers, buttes, or plains.

The usage herein of the term *sociotechnical* includes the general idea of social engineering. More specifically, it refers to the relative presence or lack of a set of conditions, thoughts, or practices in a certain locale, or at a certain time. For example, prior to 1850, before the telegraph first crossed the Mississippi River, there was little to no sociotechnical awareness, realization, or lived experience of the telegraph in the Trans-Mississippi West. The lived sociotechnical experience and awareness of the telegraph west of the Mississippi River emerged in part with the introduction and growth of the physical infrastructure of telegraphy. Yet sociotechnical experience also needed the arrival and continual presence of telegraph builders and experts. It also needed a growing awareness by local populations of the values, ills, and ambiguities of the telegraph as it became part of daily life.

Repatriation refers to the return of a person or persons, human remains, or objects and artifacts to their original location, regardless of borders and boundaries. *Extradition* is used herein in its legal sense. An example is the extradition (return) of a known or suspected criminal across national borders by law enforcement authorities of the expatriating nation to the law enforcement authorities of the nation where the supposed criminal activity took place. *Natural history* refers to older notions and combined practices of what are now usually seen as distinct fields of biology and wildlife, such as botany (plants), herpetology (amphibians and reptiles), malacology (mollusks), or ichthyology (fish). Natural historians were generalists and tended to study all these topics. Specialization also took place, and the time period studied in this book coincides with part of the long transition from natural history to specialized fields of biological study. Natural history centered on field observation and specimen collection rather than laboratory work and experimentation. Even so, some natural historians practiced horticulture and maintained nurseries, groves, farms, and estates. This was somewhat akin to laboratory work and experimentation with soil, seeds, plants, and trees.

Beginning with the title, this book often uses the word *nature*. A word with many nuances and meanings, in this study I use *nature* in more than one way. One common use of the word in this text is to invoke the phenomena of the physical world collectively. I also occasionally use the word *nature* in a different way, in the meaning that invokes the behavior, trait, or tendency of someone or something. From time to time I hint at a nuance of the meaning of *nature* that suggests the transition from something seen as novel and new to something that over time becomes seen as routine and ordinary. There is furthermore something beyond the examples just given in the ways I occasionally use the word *nature* in this book. This usage does perhaps veer toward abstraction and theorizing, but does so to aim at the human imagination of nature. In this sense, nature is something both experienced and imagined. It is something that can encompass the past, present, and future. It is something that opens the imagination of individuals to both cognitive and emotive realizations and musings. It is something to preserve, and something to shape, and something to reclaim. In this sense, nature is simultaneously a human dreamscape and a lived experience of interactions with landscapes and ecosystems. At its furthest reaches in this sense of the imaginary, nature is a continuous human dreamscape and lived experience with all of Planet Earth and all the cosmos. Nature in this way is ambiguous. It is something found in the past, present, and future of human imaginations, arrayed in an infinity of minds and ideas.

The following chapters examine, explore, and evaluate when, how, and why the telegraph wired into nature. This includes considering different geographic areas and regions of the nineteenth-century North American frontier: the Great Plains, the Subarctic and Arctic, the Pacific Northwest, the Rocky Mountains, and the Southwest. Trees, rivers, wildlife, and weather are part of the narrative. So too are surveillance, settlement, and the containment of Native Americans. The story begins on the front cover, as good a place as any to open a narrative. Henry Farny painted *The Song of the Talking Wire* in 1904.[30] In the early 1880s, Farny traveled through the Great Plains, taking notes, making sketches, gathering artifacts, and collecting stories. Farny became one of the nation's most prominent artists of the American West. Farny based *The Song of the Talking Wire* on sketches he did at Standing Rock. The story of Standing Rock and the emergent lines of technology, politics, and culture resonate across the North American frontier from the nineteenth-century telegraph to the twenty-first-century pipeline.

1

Landscapes, Ecosystems, and Prevailing Westerlies

The Great Plains

The necessity for a line of this kind has been so great that if I could only have gotten money enough out of our Army appropriations to have purchased the wire and insulators and hire a good foreman, I would have put the line up myself by the labor of the troops, and would have had it built now.

—Lieutenant General Philip Sheridan to Senator George Spencer, March 23, 1878[1]

The high ground evokes transformation. To secure and conquer the high ground was also to create the conditions to transform the surrounding terrain overseen by the high ground. Building telegraph lines, stations, and networks in western North America meant engaging the high ground literally in terms of elevation and challenging terrain. The topological, climatological, and environmental extremes of western North America far surpassed the extremes encountered in eastern North America. Securing the high ground for the telegraph in western North America also meant the telegraph needed to become a key sociotechnical infrastructure in the transformation of knowledge about the landscapes, ecosystems, and indigenous civilizations and cultures of western North America. People, trade, flora, fauna, biota, weather, and communication flowed across the continent, and people, trade, and communication had flowed and expanded across North America since the arrival of humans. Those human flows, trade patterns, and communication systems were first shaped by the landscapes and ecosystems that the people found when they first arrived. Those landscapes and ecosystems were then reshaped by subsequent generations as they lived upon and traversed the continent. To bring the telegraph to western North America meant engaging with the ongoing social and political transformations of experience and knowledge about western North America. This included exploration, white settlement, commerce, and eventual establishment of governance through territories and, perhaps someday, new states of the Union. It also meant establishing the telegraph as a new sociotechnical infrastructure in climates, ecosystems, and terrains heretofore largely unknown to telegraph builders and practitioners. For the telegraph and western North America, the high ground was multitransformational. The challenges of the high ground for the telegraph and the North American frontier during the last half of the nineteenth century were social, cultural, political, topographical, climatological, and environmental.

Much of the narrative and detail of this chapter focuses on the physical and logistical challenges faced in first building telegraph lines across the frontier. The challenges were exacerbated by encounters with an incredible range of extreme topographical, climatological, and environmental conditions. Various government reports, diaries, letters, accounts, and other writings of the era described the challenges and extremes on many occasions. Although the focus herein is on the challenges, at the same time many of those same accounts also captured the striking beauty and bounty encountered in western North America. A glorious spell of weather could set in. Sometimes game such as antelope, deer, elk, or wildfowl was in abundance, or good potable water was discovered with lakes and streams teeming with fish. From time to time edible vegetation such as wild onions or ripe berries were found and welcomed, as was discovering freshly laid eggs. Chimney Rock, red buttes, cascading rivers, flowing grasslands, snow-topped mountains, and erupting volcanoes amazed those who gazed upon them, for the scale and scope of the land was stupendous. Birds in the day and crickets in the evening sometimes sang soothing melodies. The sunrise and the sunset could be spectacular. On a clear night the sky was filled with stars, and sometimes the night sky revealed meteor showers, comets, and the stunning aurorae visible in northern latitudes. People in western North America marveled at the awesome majesty of nature and the vast infinities of the cosmos. With all this around them, they also faced constant extremes, harsh conditions, and on occasion the ultimate challenge of sheer survival. In this sense, one could experience both the ecstasy and despair of nature discussed by William Cronon, all the while one of an endless number of individuals who were, in some way or another, part of transforming the nature of western North America into a manifestation of white civilization.[2]

Crossing the Mississippi River

In 1850 white settlers in Minnesota Territory petitioned Congress to build a telegraph line across the Mississippi River, thus providing what settlers saw as a permanent connection with the federal government for assistance in territorial protection and security against Native American hostilities. The petitioners told the Senate that "Instances arise, as in cases of Indian Wars, Indian outrages, &c., when the interposition of Government, to be effective, should be immediate; yet for five months in the year, communication with and from this Territory is dependent on the ice of the Mississippi; and not unfrequently government orders and advices are from twenty to forty days reaching the seat of Territorial government." Reminding Congress that the "mere existence of a telegraph will add to the value of the public lands in this Territory," the petitioners also stated the telegraph will mean the "distant parts of this vast Republic" will be "riveted together, by unceasing and instantaneous intercourse." The settlers closed by saying they could not build their own telegraph line to connect with "Telegraphic lines South and East" because they themselves were "engaged in a contest with unreclaimed nature." They concluded that "Government is the party most directly and most largely interested in its construction."[3] Congress consistently required Native American treaties, agreements, allotments, and reservations to cede rights-of-way for telegraph lines—both commercial and military—running through Native American lands.[4] This could include additional requirements that ceded Native American property rights to water, stones, forests, game, and timber. The cessation of Native American forestry rights allowed for the acquisition of timber on Native American lands, a valuable commodity because trees that could be hewn and shaped into telegraph poles in the Great Plains and in the Southwest were relatively scarce.

The telegraph did not cross the Mississippi River into Minnesota Territory in 1850, but the telegraph arrived in St. Louis in December 1847. After initial years of intermittent success with wires strung above the river, entrepreneurs eventually found reliable signal flow to the east with a cable laid beneath the river in 1852. The early St. Louis telegraph services typically had problems with reliable signal flow, and users and operators found that "fire, sleet, hail, lightning, wind, rain, and high water" often disrupted service. In 1850 service was temporarily lost when "wires were broken twice by the large number of birds congregated upon them." The telegraph crossed the Mississippi again in 1851, this time into Iowa. New lines slowly spread up and down both banks of the Mississippi and beyond St. Louis into areas of Missouri in the early

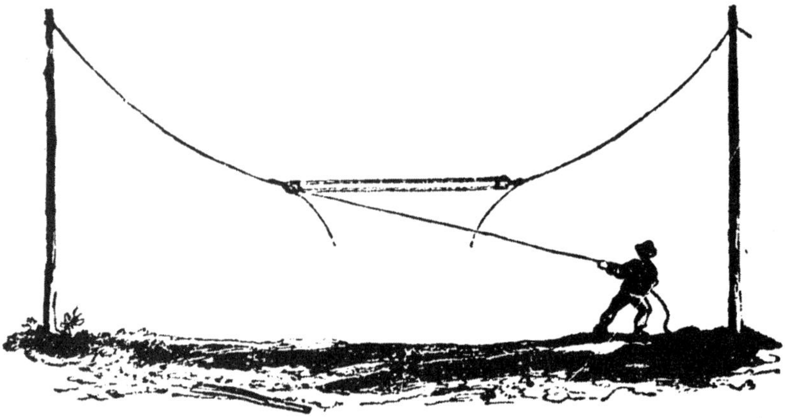

Building a telegraph line ("A Manual of Military Telegraphy for the Signal Service," p. 25)

1850s.[5] Having reached New Orleans in 1848, telegraph builders then tried to extend lines along the Gulf Coast to Galveston but were stymied by the bayous and terrain. Going to the north, lines from New Orleans reached Shreveport in 1853 and from there a line west to Marshall, Texas, began operating in 1854.[6] Securing the high ground for the telegraph in western North America involved the successful control, containment, and security of challenging terrains, such as mountains and deserts. One example is the development of the telegraph within Utah Territory (Deseret) in the 1860s.[7] Telegraph poles conveyed across the Mississippi to the American West (and telegraph wire, insulators, storage batteries for electrical power, and station equipment—the entire technical system) were, so to speak, import goods for the Great Plains and other western states and territories. They were goods that needed to be hauled westward on overland trails from the westernmost railroad depot or in some cases from a riverboat landing, or on occasion eastward from California through Rocky Mountain trails and passes. Construction and distribution centers were constantly shifting targets because of railroad construction and fluctuating river levels. Expertise in technical acumen, construction, and operation was another necessity that for the early years largely depended on individuals from eastern North America. It was no wonder that the Transcontinental Telegraph, completed in late 1861, shared much of its route with the Pony Express Route, Overland Trail, Oregon Trail, California Trail, and other trails of the territorial West. Staying close to the established trail routes (and, when possible, navigable rivers) was the only feasible way to move and install poles, string line, build telegraph stations, ship equipment, employ experts, provide territorial telegraph services, build telegraph relay stations, and maintain an operating proto-national telegraph network.

Trees, Rivers, Labor, and Technology

The early crossings of the Mississippi River by telegraph systems and the challenges they encountered evoke one of the themes of this study, the high ground. The penetration of experts and technology was not smooth and even but contingent, shaped in large part by the environments encountered during construction and operation. This leads to consideration of the landscapes and ecosystems of both eastern and western North America and how by comparison the western landscapes and ecosystems posed a distinct set of challenges, problems, needs, and conditions for the construction and establishment of telegraph networks. The landscapes and ecosystems of western North America are crucial for understanding the historical growth and particularities of telegraph lines, plans, systems, and networks. In eastern North America, technical knowledge was readily available. For trees and poles, established lumbering enterprises routinely provided hewn trees that could be hand-barked or milled into telegraph poles. Further, those trees-cum-poles could be transported from the forests to the mills and on from the mills to the markets via sluices, trails, roads, lakes, rivers, and canals serving eastern North America. They also could be transported by an emerging system of railways and post roads. Preferred timber for telegraph poles such as cedar, pine, and oak was generally available and transportable. How this played out in practice was relatively straightforward. Both the natural resources needed for telegraph networks (trees) and the manufactured equipment (wire, switches, keys, and similar) were harvested, manufactured, marketed, and transported within eastern North America with few significant environmental or distributional challenges or problems.

The situation in western North America was different. Telegraph construction in western North America during the nineteenth century was pursued by private and commercial enterprises, and also by the army. This chapter first discusses military telegraph line construction. Later, the chapter discusses commercial systems. Both military and commercial telegraph construction occurred throughout the latter decades of the nineteenth century. The army became much more active in telegraph line construction across western North America after the Civil War and with the reorganization of the Signal Corps in 1866. The activities of private enterprise and the military in building telegraph lines westward across and beyond the Mississippi River were generally distinct and at the same time often were cooperative. The military commonly took the lead in building lines needed for military activities in areas not yet served by private firms. Once poles and lines were up and a telegraph route was working reliably, shared expertise, cross-networking, interconnectivity, and other practices that served both military and commercial telegraph needs usually were the norm. The growth of railroads in western North America during the 1870s and 1880s expanded commercial telegraph networks in western North America. Railroads often built and operated (or leased) telegraph lines along their railroad rights-of-way.

In the late 1860s, the army investigated the possibility of building telegraph lines in trans-Mississippi territorial areas for controlling Native American hostilities and Mexican border raids. These plans and visions centered on Native American containment, promoting and protecting white settlers and communities, and policing the national boundary with Mexico.[8] Military construction of telegraph lines did not begin in earnest until the early 1870s, coincident with the establishment of the Signal Service, the weather-reporting branch of the Signal Corps. Signal Service traffic represented the biggest daily use of military telegraph lines in the late nineteenth century. The impetus for military lines originally came from visions of territorial military telegraph networks in western North America as an asset in the protection of white settlers and their property and communities, Native American surveillance and containment, and border control with Mexico.[9]

The lack of trees on the Great Plains and in the Southwest, the scarcity of lumbering enterprises (save small areas of California and the Pacific Coast), the absence of local equipment manufacture, a paucity of local knowledge and expertise about telegraphy, and the relative lack of navigable riverine systems, railways, and post roads all made the establishment of telegraphy in western North America an arduous challenge. An 1872 Signal Corps manual specified certain timber for telegraph posts: "red cedar or black locust . . . failing these, white cedar, spruce, white oak, chestnut, sassafras, yellow pine, or cypress" and in an emergency, "even cotton wood can be used for one year." It was best to hew and bark the trees six months before their installation as poles so the trees had time to dry and season.[10] In Dakota Territory during Signal Corps military telegraph construction in 1878, crews were advised to find local timber if possible and reminded that the "most difficult part of the undertaking is to procure the poles. Where these can be had from the timber on or near the route, they can be thence procured, though very slowly. When they cannot be had they will have to be purchased and forwarded from elsewhere."

Ideally, "cedar should be obtained if possible, failing which, other light woods with little sap, taking cottonwood only as a last resource." Hewers were told to procure "straight stems of young trees, *not less* than 20 feet long and $3^1/_2$ inches wide at top, as uniform size as possible, and barked."[11] Troops often found stands of timber and then hewed and hauled poles, and on other occasions the work was done by local farmers, ranchers, white settlers, or other local suppliers. Such suppliers were not always averse to price-gouging. In 1876 Lieutenant Philip Reade chastised pole suppliers in Arizona Territory: "When fine pine poles, hauled forty miles in New Mexico, can be purchased for half a dollar each, it seems strange that Arizona dealers should charge four times that much for inferior cottonwood."[12] In Nebraska Territory, a rancher named Jack Morrow controlled access in the early 1860s to a large stand of cedar timber, which he would not let anyone else cut. Morrow cut and sold "two thousand fine cedar telegraph poles" from his grove for use on the Transcontinental Telegraph.[13]

Despite Signal Corps guidelines and precautions, the 1878 Dakota military telegraph lines had tree-to-pole problems. In some sections, the poles used were "generally small and fully one third are dead wood in many cases worm eaten and already in incipient stages of dry rot. The distance from which poles were hauled however explains why such poles were used."[14] Hauling poles was, to put it mildly, brawny work, especially if wagons pulled by draft animals such as oxen could not reach an area where trees were felled and the trees-cum-poles therefore needed to be hauled away by hand. When a portion of military telegraph line was built in Arizona Territory near the Gila River in 1875, poles needed to be hand-carried from supply wagons by construction crews as far as eight miles.[15] A minimum of five men and one foreman could haul by hand two or three lighter-weight seasoned poles per load, but for two to three "green posts of oak, locust, or chestnut, ten men will be needed." Pole haulers were expected to match the pace of post-hole diggers, itself no calming chore. A team of six to ten men working in pairs were expected to dig by hand four to five miles of post holes each day, seventy-five yards apart, assuming "clay or loam in which to make the holes." All the while out front of the tree-hauling and hole-digging crews were "one or two ax-men to clear the way" of trees, shrubs, and undergrowth. Clearance around the route of the telegraph line reduced the likelihood of damage to the line caused by falling trees or by brush fires. At the end of the construction team was the wire party, a six-man crew with four climbers, one to a pole, four poles strung at a time. A wire reel was part of the crew, ideally carried on a wagon and, if not, the reel had to be hauled on the back of a mule or by

Telegraph construction, overland stage, and rider at Red Buttes, Wyoming (William Henry Jackson Collection, Scotts Bluff National Monument, National Park Service, catalog number SCBL_148)

hand.[16] Poles erected from green trees were more susceptible to splitting as they seasoned and to early stump rot.[17]

Finding and harvesting suitable trees for telegraph poles, telegraph stations, cabins, forts, agencies, other structures, objects, items, and activities such as building a cooking fire was a continual challenge in much of western North America. Hervey Johnson, a Union Army soldier stationed in the Great Plains during and after the Civil War, wrote his family that harvesting timber from mountainsides and deep gorges was brutal for men and for draft animals: "it is killing work on men and oxen."[18] Oxen and wagons for the wire and other equipment were always needed because "six miles of wire, insulators, and nails weigh one ton."[19] Regular army troops usually carried out the construction of military telegraph lines under the supervision and guidance of Signal Corps officers and personnel, although the army hired private contractors on a few occasions. Wagons transported poles, insulators, and spools of wire as near to the construction site as possible. If, however, pasturage was not available in or near the construction area, then forage also had to be hauled for horses and draft animals, and of course provisions for construction teams needed to be hauled to work sites. Sandstorms and snowstorms could wreak havoc. In the 1870s a military telegraph line built across sand dunes between San Diego and Yuma occasionally had its poles buried, toppled, or uprooted by shifting wind-borne sand.[20]

As the Signal Service expanded its mission to include monitoring river levels, other challenges emerged. The weather observer at Fort Scully, Dakota Territory, reported in 1873 that the river gauge on the Missouri River was about 1½ miles from the fort, and the twice-daily measurement "exposes the observer to a twofold danger—the rifle and arrow of the Indian, or drowning."[21] The Missouri River itself was a challenge, with driftwood snagging at connecting cables on the river bed and the need for stringing wire eighty feet or more above the river at river crossings to avoid potential tangles with river ships and the smokestacks of steamboats.[22] Other rivers, such as the Platte, were for all intents and purposes unusable for shipping telegraph equipment and for floating poles because of their shallow depth. Many stretches of the Platte were too shallow to navigate even in a canoe.[23] A significant river flood could change the course of a river, which then might necessitate rerouting a telegraph line. Flooding on the Gila and Little Colorado rivers in 1891 washed away riverbanks and changed the courses of those rivers in ways that stymied telegraph repair parties.[24] River crossings often proved to be a weak link for a telegraph route. When a line crossing a river went down, messages often were shuttled across the river by boat until the broken line could be reattached on the two sides of the river. If the weather and winds were favorable during telegraph line construction or repair, wires were flown across rivers by kites and then strung on poles.[25] Winter was a continual challenge for many telegraph stations in western North America. The Pembina Signal Service station struggled with "irregularity" in transmitting weather observations to Washington during the 1870s because of the "unprotected and exposed condition of the line."[26] After the completion of the Transcontinental Telegraph in October 1861, storms often caused damage to the line throughout the seasons. A severe storm in April 1864 took down many poles in an area of Idaho Territory, disrupting the line for three days. A bolt of lightning in Colorado Territory near Julesburg in summer 1864 "struck one of the telegraph poles not far from us, and splintered the poles or damaged them for a great distance on each side." Such events were not unusual.[27]

On occasion, there was positive news to report on trees and poles. A section of military telegraph line constructed in western Texas in 1875 went well. "The line between Fort Richardson and Fort Griffin is 69 and 16/25 miles long, and is constructed entirely of cedar poles."[28] The town of Telegraph, Texas, now a ghost town, was named because nearby cedar groves provided timber for telegraph poles.[29] Even with these cases, the good news in Texas was somewhat muted. The challenging logistics and strenuous physical labor of transporting cedar poles in some areas of Texas often meant that installation could take place only three days a week. In West Texas, the Great Plains, and much of the Southwest, the need to transport forage such as hay or shelled corn for horses and draft animals was typical. Throughout Texas and the Southwest, it was routine to replace transported cedar poles with locally available timber such as post oak, red oak, pine, pecan, hackberry, walnut, ash, and red elm, interspersed as possible with red or white cedar. Cottonwood and willow were used as a last resort. At one point, military telegraph construction work near the Rio Grande was delayed by several weeks because of problems in shipping poles from Virginia.[30]

Climates and Landscapes

Many of the poles supplied for the 1865–67 Western Union Telegraph Expedition in Alaska, Siberia, and Canada were cut in Puget Sound and shipped to both Alaska and Siberia and then barked and shaped on site.[31] In the 1870s Western Union also offered fir trees from the Puget Sound area for construction of the military telegraph line from San Diego to Arizona Territory, although the army ended up purchasing seasoned California redwood poles from them.[32] Members of the Western Union Telegraph Expedition survey team in Siberia found stretches of land "entirely bare of trees" and "ill adapted" for telegraph lines and other areas where timber was of "inferior quality (but) would furnish poles sufficient."[33] Standing timber in Siberia was a mixed lot: "some land wooded, some tundra, prone to floods"[34] but another locale "so abundant with timber" that it would be possible to cut poles "on the precise spot where they are desired to stand."[35] Abundant timber did not solve everything, as deep snow often hindered the felling of trees.[36] The hindrance that deep snow caused for hewing and hauling timber also occurred in other places, such as Dakota Territory, where the timber harvest in winter was "awful work" with "snow three feet deep" and "the timber all being down is sometimes hard to find."[37] Heavy snows in the Sierras brought down some early commercial telegraph lines in California.[38] The snow weighed heavily on the telegraph line leading to the Signal Service weather observation station at the summit of Pike's Peak in the 1870s and in "winter the snow broke down the poles so repeatedly that at last orders were issued to take the wire from them, above timber-line, and place it on the snow. After this was done there were but few interruptions." The rigors of the Pike's Peak Signal

Clearing the way for a telegraph line in Alaska, early 1900s (Archives of the University of Alaska–Fairbanks, and the Signal Corps Report collection, identifier UAF-1996-3-4)

Service station were such that "a year's tour of duty is sufficient even for the most enthusiastic student of meteorology."[39] Frost and ice on telegraph lines at and near the summit of Pike's Peak could collect and build up "nearly to the thickness of one's arm."[40] Similar extremes also occurred at the Signal Service station atop Mount Washington, New Hampshire, which along with some other stations began to use the telephone in the 1880s. The Mount Washington station reported the "benefits derived from the telephone have been great, as we use it as a test for our mountain wire, thereby saving many unnecessary trips down the mountain."[41]

When Frank Pope explored western British Columbia for the Western Union Telegraph Expedition, he thoroughly considered microclimates and local landscapes. Telegraph lines would do best, Pope concluded, when run along the lee side of a river or lakeshore that had the least amount of tree growth, as the lines on the lee side would likely be less subject to damage by wind-toppled trees. Doing so was not always easy with wide rivers that meandered, as the lee and windward sides of a meandering river could change with each bend, necessitating occasional river crossings with telegraph lines built above the water. Rainforests were to be avoided, despite an abundance of timber, because of the dense undergrowth, toppling trees, ravines, challenges in building roads and hauling materials, and difficulties of maintenance. In general, it was best to avoid "the woody and rainy country" and seek areas with less annual precipitation, even if it meant fewer trees in the area, thus necessitating the hewing and hauling of timber from elsewhere. Timber was abundant in some areas and sparse in others, but Pope argued the abundance of local timber near proposed routes should not be the sole factor in routing lines.[42] When a Signal Service weather observation station opened in Olympia in Washington Territory in 1877, telegraph reports from Olympia often were delayed because the branch line to Olympia was "frequently prostrated by falling trees."[43] In the 1880s in the Cascade Mountains, a forty-mile stretch of military telegraph line was attached directly to standing timber. The swaying of the trees during wind and rainstorms caused frequent breaks in the line.[44]

The challenges of trees, poles, and local landscapes and ecosystems always remained. By the 1880s the Signal Corps was on occasion dispensing with trees and poles altogether, turning to iron poles and masts whenever feasible. General William Babcock Hazen, Signal Corps Commander in the 1880s, reported that trees and poles could not always survive the range of environmental and climatological conditions of western North America, explaining that "fires often run over the country, and in the dry seasons wooden poles are destroyed. This makes it necessary to use iron poles, and to secure them against rust they must be galvanized. These, at that time, could only be bought in the East, and at very few places even there."[45] Iron poles and masts were increasingly used in the 1880s throughout the Signal Corps military telegraph system, often in the repair or rebuilding of older lines. This included areas prone to prairie fires, such as the Great Plains, and also included areas where wooden poles proved susceptible to arid conditions, such as Arizona and New Mexico Territories.[46] The endless problems with cottonwood in the Dakotas led to plans to install iron masts for Dakota military telegraph lines in 1887.[47] Maintaining a military telegraph line in the area of Fort Maginnis and the Poplar River in Montana Territory was prohibitively expensive because the surrounding area was "wild and desolate in the extreme, and generally without inhabitants, so that at Galpin, the repair station" the Signal Service needed to station two men "since one will not live there alone."[48] Iron masts also replaced wooden poles around Fort Bridger in Wyoming Territory in 1889 and 1890.[49] Iron masts were a vast improvement for dealing with the landscapes and ecosystems of western North America but not a panacea. Iron masts near Brownsville, Texas, and the Rio Grande washed out and became unmoored in 1888 during floods and heavy storms. The break in communication caused a problem, as the military telegraph line bolstered policing and border security in "a section of country much infested by outlaws and all classes of criminals."[50] In the latter 1880s military telegraph construction in western North America declined. The continual emergence and expansion of commercial telegraph services in a region or territory was seen as preferable to maintaining military telegraph lines wherever feasible, particularly whenever a railroad installed and operated telegraph lines along its own right-of-way.

The railroad, a classic symbol of settlement and expansion on the western frontier, proved an asset to the Signal Corps, the Signal Service, and their telegraphic pursuit of weather data from across the continent. Another classic symbol of settlement and expansion—the fence—was a different matter. In the 1880s the Signal Corps routinely began to reroute its own military telegraph lines in Washington and Idaho Territories to abandon fenced-in land, as repairs and maintenance in fenced-in areas proved difficult.[51] Settlement, ranching, and the local climate and ecosystem sometimes presented a tripartite challenge. The Signal Corps personnel maintaining military telegraph lines between Fort Elliott and Fort Sill in Indian Territory (Oklahoma) in 1883 reported that line was "liable to many interruptions from prairie fires, thunder-storms, floods, and malicious acts of cowboys and outlaws" but kept the line generally operable. Other ecosystem challenges in the area for telegraph lines included river washouts and quicksand.[52] In Nebraska Territory alkali flats, lakes, and waterholes were exceedingly dangerous areas of entrapping mud "in which many an ox and buffalo have perished." One Union Army soldier guarding telegraph lines near alkali flats recalled almost "getting into one myself one night."[53]

Wildlife, Draft Animals, Insects, and Diseases

On occasion, the environmental and climatological tasks facing the Signal Corps and Signal Service might have seemed to the officers and troops almost epic. In 1877 the Signal Service began a cooperative effort with the United

States Entomological Division "in collecting and summarizing all obtainable data relating to the origin, flight, habits, and devastation of the grasshoppers or locusts in the regions west of the Mississippi River." Reporting with relief that recent seasons had seen no significant infestations, Hazen confirmed that his telegraph network stood prepared if future swarms ever meant that "opportunities may offer for this service to give notice, some time in advance, of the approach of these destructive insects."[54] Although locusts and other destructive insects did not usually pose a significant threat to the physical infrastructure of trans-Mississippi telegraph lines, they often were a nuisance to construction crews. Mosquitoes and biting flies abounded, and an 1849–50 yellow fever epidemic temporarily halted telegraph progress in Missouri, as did a cholera outbreak in 1850. Cougars, wolves, bears, and poisonous reptiles also could be hazardous to crews. A litter of skunks taking up residence at a construction site, encampment, or building brought along many moments of pungent misery.[55] Lines hung with too much slack or hanging too low were from time to time broken by cattle herds, particularly on open grazing land in the Southwest.[56] On one occasion, slaughtered cattle actually helped keep a telegraph line up and running. A scarcity of glass insulators in the Confederacy during the Civil War led one hard-pressed telegrapher near Shreveport to replace glass insulators with cow horns.[57] Glass insulators emerged as the insulator of choice by the early 1860s, having been preceded by an array of so-so to dysfunctional products variously using (alone or in combination) wood, pine pitch, beeswax, cloth, gum shellac, hooks, rubber, iron, and other handy odds and ends. Some of these early hodgepodge in-

Bear confronts man (William Henry Jackson Collection, Scotts Bluff National Monument, National Park Service, catalog number SCBL_73)

sulators occasionally worked better and lasted longer if they were regularly "greased," which added to labor and maintenance.[58]

One animal of western North America in particular, for a short time, presented a significant challenge to the physical infrastructure of telegraph lines in the Great Plains: bison. Bison were known to rub up against and knock over telegraph poles, and a migrating herd in sufficient strength, speed, and numbers could easily topple many miles of telegraph poles and line.[59] In terms of significant damage to the physical infrastructure of a telegraph line, there was no equivalent in eastern North America, save perhaps the passenger pigeon. When gathered in great numbers, passenger pigeons often broke the limbs and even split the trunks of trees in which they alit, and they could break telegraph lines when alighting in a large roost. By the early 1870s, both the bison and the passenger pigeon were unceasingly slaughtered in huge numbers by white settlers, bird netters, and railroad hunters wherever and whenever they were found. Indeed, the telegraph aided, abetted, and accelerated those slaughters, giving immediate word of the arrival and presence of large gatherings of bison and pigeons in particular areas. The railroads (and also Great Lakes shipping for the passenger pigeon) brought the hunters, harvesters, suppliers, armaments, buffalo skinners, pigeon netters, pigeon clubbers, and barrels for the pigeon packers by the thousands. As the slaughters progressed, the railroad and telegraph also coordinated the shipping and distribution of huge heaps of buffalo skins and hundreds of barrels stuffed with pigeon carcasses. For example, local newspaper reports of the 1871 Kilbourn City, Wisconsin, and 1878 Petoskey, Michigan, passenger pigeon nestings traveled quickly to newspapers in Chicago, Milwaukee, Detroit, Cleveland, Buffalo, and elsewhere by telegraph and the news soon brought hunters and netters in abundance via roads, ships, and railways.[60]

In addition to animals, insects, and fish native to the landscapes and ecosystems encountered by telegraph construction crews in western North America, draft animals were routinely present. Oxen were usually preferred for hauling construction wagons, although oxen could not easily reach all areas of telegraph construction. Horses and mules were familiar domestic and draft animals for construction crews across much of the Trans-Mississippi West. Dogs often joined these crews as companions and also were sometimes used in guarding camps and on hunting forays. In Alaska and other Arctic regions, dogs were crucial as draft animals, and dogsleds were a mainstay during the first explorations and attempts to build telegraph lines in Alaska and other Arctic regions in the nineteenth century. In most cases, the sled dogs were a mix of breeds. Robert Kennicott described the dogs from his

Signal Corps mushers and dog teams, Alaska, early 1900s (Archives of the University of Alaska–Fairbanks, and the Signal Corps Report collection, identifier UAF-1996-3-8)

1859–62 subarctic and arctic travels as a collection of purebred "Yukon dogs" with some other "Yukons" interbred with Indian dogs known as "geddies." Kennicott was fond of his sled dogs and admired their tenacity and stamina. He thought his lead dog Nogah was "as hard-headed as a pig."[61] Crews usually did not ride on the dogsled but ran along beside the sled. Running alongside a moving dogsled could be exhausting, particularly in deep snow (if one had not yet mastered snowshoes) or high winds. One could ride on a dogsled for a short period without hazard. To ride on the dogsled for too long presented a significant risk of falling asleep and then freezing to death.

The nutritional needs of the dogs also presented challenges. Unlike foraging draft animals such as mules or oxen, dogs needed huge amounts of protein. If fish were abundant and could be caught by crews or procured from local indigenous populations, dog feeding went well. If game was abundant and deer or bear were procured, the dogs ate the same protein-providing foods as the construction crews, and construction crews also ate fish. If protein was scarce—or for that matter, even if it was not—sled dogs would still eat almost anything. A crew that failed to keep the sled dogs securely separated from the dogsled with its leather ties, separated from boots and rucksacks, separated from anything made of leather or animal skins, and away from the food supply of the construction crew had a big problem on its hands. The dogs would ravenously consume anything they could find. By the end of the nineteenth century, domesticated reindeer herds were introduced into Alaska. Capable of foraging on lichens, mosses, and other plant life of the arctic ecosystem, telegraph construction crews sometimes used reindeer as draft animals, draft animals that could feed themselves on available forage. Reindeer were less susceptible to injuries from crusty snow and ice (which cut the legs of mules) and also provided meat and milk. By 1900 military telegraph construction crews in Alaska were using reindeer to haul poles and equipment to telegraph construction areas.[62]

Reduction of Military Telegraph Construction

By the latter 1880s, the Signal Corps saw overall reductions in the miles of its military telegraph lines as commercial services expanded. Even with the growth of commercial enterprises, trees and poles for military telegraph lines always needed attention. Poles continued to rot in areas of Montana Territory. Iron poles replaced wood poles in Kansas and were deployed throughout Arizona and New Mexico Territories. As opposed to drought and arid conditions, the coastal rainforest climates of northern California, Oregon, and Washington Territory presented different challenges for maintaining military telegraph lines, in part due to forest fires, falling timber, and mudslides. In Idaho Territory, good pine and cedar stands for telegraph poles could still be found in the late 1880s.[63] In the late 1880s and into the 1890s, Signal Corps construction of military telegraph lines steadily decreased as commercial telegraph services in western North America became more viable at local levels and forts and stockades were closed. A new railroad line invariably brought a new commercial telegraph line. The closing of a fort sometimes exacerbated problems on military telegraph line maintenance and repair, as there was no longer a garrison of soldiers nearby to serve as a repair team.[64] Repairs, upgrades, maintenance, and occasional new line construction often was realized by salvaging lines, poles, and iron masts from recently abandoned lines for construction in new areas. Some of the new lines of the 1880s and 1890s were short-haul military lines aimed at making better connectivity with the growing commercial telegraph lines and networks.[65] Signal Corps lines on seacoasts over time were transferred to the Coast Guard, with the last lines transferred in 1929.[66] The biggest factor in the reduction of military telegraph lines in the 1890s and early 1900s, however, was the expansion of commercial telegraph services. In 1888 the Signal Corps reported it had "built in the last fourteen years about 8,250 miles of *military* telegraph lines, which fortunately the rapid advance of settlement and civilization has reduced to less than 2,000 miles."[67]

The need for constructing military telegraph lines may have abated in the last decades of the nineteenth century. Yet the perceived need for more Signal

Service weather observation stations beyond the Mississippi and throughout the North American frontier, networked by both military and commercial telegraph systems into the nationwide Signal Service data collection network, had increased. In an 1884 report, the Signal Service noted that its network extended across the continent from the Atlantic to the Pacific Ocean as well as northward from the West Indies and Gulf of Mexico to the Canadian frontier. Nevertheless, they were "still in need of more stations in the interior of the country" as well as in the Northwest Territories of Canada to better collect weather observations. More observation stations would increase confidence that forecasts and results were built upon "the solid ground of nature." The "*extreme northwestern* frontier is, perhaps, the most important task," and expansion of telegraph lines and Signal Service weather observation stations meant "fresh stimulus to almost every meteorological investigation which the service is now pushing."[68] An 1880 Signal Service pamphlet explained that the observation stations across the Mississippi in the northwestern areas of the continent were the "vital arm" of the Signal Service. These stations observed and measured "the heart of the American storm-field—whence ceaselessly issue those pulsations of atmospheric energy that are felt from the 95th meridian to the Atlantic and from the Lakes to the Gulf" and in so doing contributed valuable data to weather forecasts for the entire nation.[69]

Growth of Commercial Telegraph Routes

Military telegraph lines and the work of the Signal Corps were fundamental to the growth and development of telegraph lines beyond the Mississippi. Growth and development also emerged from the efforts of private and commercial interests. By the end of the nineteenth century, commercially operated telegraph lines and networks built by telegraph companies, railroads utilizing their right-of-way for telegraph lines, telephone companies, and other entities had progressed to the point where military telegraph lines were less fundamental in western North America. In the 1880s and 1890s, the Signal Corps consistently supported such private enterprise and, when those commercial telegraph lines were sufficiently reliable, usually moved their signal flow away from their own military lines and onto commercial lines. The transfer of traffic also meant a reduction in construction, maintenance, and pole replacement for the Signal Corps. Both the military and commercial telegraph lines of western North America were central to the conquest and transformation of the high ground.

The eventual success and profitability of commercial telegraph lines in western North America was, at the start of the 1850s, doubtful to virtually all observers. "Only the most ambitious—or foolhardy—entrepreneurs envisioned constructing a line" across vast and barren stretches of the Trans-Mississippi West but "when it came to the telegraph industry, such men were in abundance."[70] Abraham Lincoln told Hiram Sibley his proposal for the Transcontinental Telegraph was a "wild scheme" in the face of the problems of getting poles and equipment into place on the treeless Great Plains. Lincoln also saw security problems, concerned that "as fast as you build the line the Indians will cut it down."[71] Small-scale and local private and commercial telegraph line construction emerged in scattered places across the Trans-Mississippi West in the early 1850s. Despite the clear utility of telegraph service in California and the Pacific Coast in the early 1850s, telegraph construction at first was "taken up with more seeming reluctance."[72] Miners and prospectors in Nevada and California connected some communities and camps by telegraph in 1853. Marysville, San Francisco, San Jose, Stockton, and Sacramento were all linked together by telegraph lines that year. The Marysville construction crew at various points became lost in local fog.[73] Attempts in 1856 to lay cable beneath the waters of the San Francisco bay area produced unsatisfactory results and, as in other areas of water at that time, underwater cable attempts were abandoned. Traditional pole-strung lines were built around the perimeters of the bay in 1856. Various mergers and combinations of commercial telegraph companies in California during the 1850s eventually led to the formation of the California State Telegraph Company, which, along with Western Union, played a major role in the eventual completion of the Transcontinental Telegraph in October 1861.[74] During the 1860s, the telegraph grew south in California. Lines from San Francisco reached Los Angeles by 1860 and finally reached San Diego by 1870.[75]

Activity to link Oregon and California telegraphically began in Portland in 1862 and connected with California by the end of 1863. Earlier efforts in Oregon had begun in 1854 in the Willamette Valley with a line from Portland to Corvallis, completed in 1856. The wire for that line, however, was strung directly on trees and as a result was easily damaged by wind, storms, and falling timber. The environmental and climatological damage to the line was compounded by a relative lack of routine maintenance by the builders and operators. Throughout western North America in the 1850s, it was not unusual for early lines to fall into disrepair or disuse for lack of maintenance, leading to the "disgust of towns" served by those lines and bitter complaints from local investors.[76] Eventually the Willamette line was abandoned, and Oregon telegraph efforts turned instead toward interconnection not only with California, but also with Vancouver in Washington Territory and Victoria, British Columbia, both reached in 1864.[77]

Sufficient and suitable timber for poles posed a problem, particularly in connecting California with the East. In Utah Territory, the Mormon community provided timber for many miles of telegraph lines, in part utilizing "stunting pine and quaking asp" known by local Mormon communities to be growing in canyons and gorges. Mormon efforts at securing trees and poles were intermittently successful and on occasion construction crews had to travel significant distances to hew and haul trees for poles.[78] The passage of the Transcontinental Telegraph through Salt Lake City later spurred the growth of telegraphy within Deseret and its Mormon community after the Civil War. A telegraph training school was established, and funds were either granted by the Church of Latter-Day Saints or locally raised for wire and other manufactured supplies. Local communities cut and hauled timber for poles, dug the postholes, and erected lines.[79] When the Transcontinental Telegraph line was completed, Brigham Young was honored with an opportunity to be among its first users. His telegram expressed thanks "in giving me privilege of first message to California" and in a prescient moment that revealed he knew of larger ambitions, urged the Californians to "join your wires with the Russian Empire, and we will converse with Europe."[80] In Wyoming Territory, some miners and prospectors temporarily abandoned their mineral claims in order to earn money delivering telegraph poles for the Transcontinental Telegraph.[81]

Wyoming Territory had several stations along the Transcontinental Telegraph, but Colorado Territory had only Julesburg. Neither Denver nor the Colorado gold fields were immediately connected as stations directly on the Transcontinental Telegraph or connected with branch lines, perhaps because of a lack of local funds.[82] In Missouri and Arkansas, the establishment of the Butterfield Overland Mail Route influenced some early telegraph construction. In the 1850s, one line was run along part of the northern branch of the trail, from St. Louis to Jefferson City and beyond the trail to Boonville, eventually extending to Kansas City and crossing into Kansas in December 1858 and on to Leavenworth in January 1859.[83] Some of this route dispensed with poles and attached line directly to standing timber. Efforts began in 1859 to extend the line along the southern branch of the trail to Fort Smith, Arkansas, while line construction was begun in an eastward direction from Los Angeles. Securing and distributing poles for these endeavors was, without surprising, a constant challenge.[84] One newspaper editor in Arkansas writing in early 1860 bemoaned the status of Arkansas as "the only black spot in the 'telegraph galaxy of the nation'" because of its lack of any operating telegraph routes and lines. Poles to Fayetteville and a line to St. Louis were completed by May 1860, but the first messages between Fayetteville and St. Louis were celebratory telegrams sent in July to celebrate Independence Day.[85]

Wagon train, trail, and telegraph line (William Henry Jackson Collection, Scotts Bluff National Monument, National Park Service, catalog number SCBL_3126)

The Transcontinental Telegraph

The most significant telegraph project of western North America prior to the Civil War was the Transcontinental Telegraph. Largely organized and built by Western Union in coalition with California telegraph companies, Mormons, and other individuals west of the Mississippi, financing and plans, including congressional support, were in place by 1860.[86] Working east from Sacramento and west from Omaha—keeping in mind these were constantly shifting organization sites and termini as construction proceeded in both directions—the Transcontinental Telegraph took advantage of some already-established telegraph routes as well as established overland trails. The route also had similarities with the Pony Express route, and as construction progressed, the Pony Express covered a slowly shrinking route while riders carried messages between the nearest operating telegraph stations over the sections of the telegraph route under construction but not yet completed.[87] As construction on the Transcontinental Telegraph proceeded, the combination of Pony Express and telegraph relay reduced the time for news and messages to travel eastward from San Francisco to about ten days, although occasional telegraph problems could extend this duration. One such instance occurred in Kansas in 1860 when horse thieves severed telegraph lines around Leavenworth.[88]

Edward Creighton played a major role in surveying and overseeing construction for much of the line westward from Omaha to Salt Lake City, and James Gamble did the same eastward from California, although they were only two of many individuals involved in surveying and construction.[89] Creighton had first built telegraph lines in Ohio and New York in the 1840s and 1850s, particularly a line west from Buffalo that later became a major trunk for Western Union. Moving to Omaha in the 1850s, Creighton found local interest in the telegraph was widespread. The goal for Omaha and Nebraska Territory was the West, particularly California and the gold fields. The *Omaha Times* asked in 1858 when the Transcontinental Telegraph would finally be built, noting that "We have been talking about a telegraph to the Pacific Coast ever since California was organized."[90] Omaha was not the only city eager to reach California by telegraph lines. The call to push on to California was heard in many locations. As soon as the first telegraph lines arrived in St. Louis, local newspapers extolled the telegraph builders to "carry it forward to the Pacific Ocean."[91] Senator William Gwin of California was one of many in the 1850s who believed the establishment of some sort of national road or trail from the Mississippi River (or thereabouts) to California would also ensure a national telegraph line.[92]

Countless tons of telegraph wire on reels, insulators, carboys, instruments, kegs of nails, shovels and axes, and other equipment moved westward from Omaha to build the Transcontinental Telegraph. "Nearly one thousand oxen were found necessary" for transporting "camps, food, wire, and poles."[93] Supplies arrived at Omaha for the most part via the Missouri River, "Omaha [being] a great steamboat town."[94] Many wagons pulled by oxen or mules left Omaha with three or more tons of telegraph supplies.[95] To no surprise, securing poles was vexing, particularly around Julesburg, the station planned for Colorado Territory. Groves of red cedar were found in Cottonwood Springs, one hundred miles to the east and were harvested despite being "very crooked and unsightly" and then used for much of the line between Julesburg and Fort Laramie. Trees also were necessary for building telegraph stations, typically made out of logs or, if need be, sod. Forts, stockades, trading posts, agencies, ranches, white settlers, trail travelers, and Native Americans all used trees for many purposes. Trees were a scarce commodity for everyone, not just telegraph construction crews.[96] A telegraph station also needed a local water supply or well, as did stations on the stage routes. A telegraph construction team working westward from Omaha encountered a stagecoach company digging a well west of Julesburg in July 1861. The crew was to dig to one hundred feet and "then, if no water signs are found, to abandon the work. . . . No water was found; no station erected." As the telegraph progressed, telegrams were sometimes sent along working sections of the Transcontinental Telegraph to determine whether an open seat might be found on part of a stage route.[97]

Upon completion of the Transcontinental Telegraph in October 1861, the line went into regular operation and immediately proved profitable. More than two hundred messages were sent on the Transcontinental Telegraph in its first full day of operation, and messages continued to be sent "as fast as the operators could transmit them."[98] Hundreds more telegrams traversed the wire within its first week of operation, leading Jeptha Wade to comment "it will soon be necessary to put up one or two more wires to meet the urgent wants of the commercial public and the Press."[99] Success meant the end of the Pony Express, although "in its death it enjoyed the honor of giving way to one mightier than itself, a medium which could do in minutes what it took days to accomplish with horseflesh."[100] The *St. Joseph Weekly Free Democrat* waxed eloquent in October 1860 on the wonder of it all, even before the completion of coast-to-coast telegraph connectivity. Hailing the eventual end of isolation and the continual conquest of nature, its editors believed the new telegraph line "unrolled to us the great American Panorama; allowed us to glance at the future home of one hundred million people; and has put a girdle around the earth in forty minutes."[101] Among the many telegrams back and forth between California and points east were money transfers, an important development because of the wealth of California and its distance from eastern North America. Gold and silver strikes, first in California and later elsewhere, also stimulated telegraph traffic from financial speculators.[102] The great project completed, telegrams now flowed apace across the full spread of the nation. Yet the Transcontinental Telegraph remained vulnerable to local ecosystems and climates, indigenous peoples, white settlers, and the escalating political tensions of the Civil War. "Nature assaulted the line with snow, floods, and high winds; the Indians outgrew their early fears, and they, and renegades dressing as Indians, pulled down the wire and burned the poles; emigrants used the poles for firewood; buffaloes wore them away by their persistent rubbing; and in Missouri Confederate raiding parties systematically destroyed the line."[103]

War on the Horizon

The escalating tensions of the Civil War and opening of hostilities were coincident with the construction of the Transcontinental Telegraph. Already subject to occasional vandalism and sabotage by Confederate sympathizers upon its completion in October 1861, the Transcontinental Telegraph added telegraph branch lines and a new section of the main route. By the first few months of 1862, a more northerly and more secure route for the Transcontinental Telegraph replaced the original route lines that ran through Missouri from St. Louis to Omaha via Union, Boonville, Lexington, and St. Joseph. The new northern route replaced St. Louis with Chicago as the eastern terminus. The rerouting also meant several cities in Iowa, such as Cedar Rapids and Des Moines, would for the first time enjoy telegraph service and networks.[104]

Built for a multitude of reasons, the Transcontinental Telegraph also was a delayed response to the outcome of the war the United States had waged with Mexico from 1846 to 1848. The 1848 Treaty of Guadalupe Hidalgo ceded vast territories to the United States, including California. Signed on February 2, 1848, the treaty was consummated coincident with another event: the discovery of gold in California at Sutter's Mill on January 24, 1848. The California gold strike was completely unknown to the treaty signatories. Despite efforts by those who found the gold at Sutter's Mill to keep the strike secret, news of gold soon hit the streets of Sacramento and San Francisco. Word of the California gold strike spread across the United States and around the world. President James Polk officially notified Congress of the gold strike in December of that year.[105] California quickly transitioned from territorial status and became a state in 1850. A new state was part of the Union, and that state happened to be the location of the greatest gold deposits ever found, anywhere, up to that time. Communication to and from California with the rest of the nation was slow. Two communication routes were by ship around Cape Horn and by ship along the coasts and then overland across the isthmus of Central America. The third communication route was on overland trails. The telegraph had barely crossed the Mississippi River by 1850. As Robert Luther Thompson remarked about this situation of long-distance transportation and communication, "the bonds which united the new state with the central government in Washington were more sentimental than real."[106] California was isolated.

As the telegraph in the 1850s grew hopscotch, piecemeal, and catawampus to be scattered here and there beyond the Mississippi River, taking root across the landscapes and ecosystems of the Trans-Mississippi West, a young army assistant surgeon arrived at Corpus Christi, Texas, in November 1854. From there Lieutenant Albert Myer began to make his way from the docks of Corpus Christi to his post at Fort Duncan, Texas.[107] In the years to come, Myer would develop an innovative system of signal communication based on flag signaling for the army and would first test his system in 1859 during army campaigns against the Navajo. He would later go on to found the Army Signal Corps, be removed from command of the Signal Corps during the Civil War, and be reinstated to that command in 1866. Myer would found the Signal Service as a weather observation branch of the Signal Corps in 1870, and in that post he would oversee thousands of miles of military telegraph-line construction in the Southwest and elsewhere. He would develop strategies and tactics for the use of the military telegraph as a central asset for the army during Native American surveillance, containment, and conflict. None of those thoughts occupied Myer as he made his way to Fort Duncan in 1854. He thought instead about his travels through a harsh and barren landscape, a brutally hot climate, and an ecosystem virtually devoid of edible plants and potable water. The outbreak of the Civil War, however, brought Myer back to Washington to command the Signal Corps for the Union Army. Unlike the war with Mexico, the telegraph was clearly going to play a significant role, some way or another, in the Civil War. Back in eastern North America, Myer prepared to take part in the greatest existential crisis that the nation had faced. Yet he never forgot his first encounters with western landscapes and ecosystems. For Myer, the Civil War meant that the challenges of the high ground and the North American frontier would have to wait.

2

Storms Moving in a Ring of Fire

The Civil War

> How striking that in the darkest days of the Republic's history, while a battle was drawn between North and South, this line was fashioned to bind East and West.
>
> —P. Raymond Nielson, "Edward Creighton and the Pacific Telegraph"[1]

If any significant security measures existed for the telegraph within the District of Columbia—indeed for connectivity to any telegraph line prior to April 1861 and the onset of the Civil War—they are not strongly evident. Various codes and ciphers were used in telegraph messages for Washington and elsewhere, but they often consisted of condensed words for the purpose of reducing the cost of a telegram rather than securing its contents. As the rebellion commenced in the middle of April 1861, telegraphy in Washington strained from the stress to the near breaking point. At various moments, the city was either swamped with telegrams or bereft of telegraphic connectivity. All telegrams at the time passed through commercial offices in Washington. For all intents and purposes, no federal building had direct connectivity with its own government-run telegraph station, meaning the federal government had to use messengers to shuttle telegrams from government offices to public, commercially operated telegraph stations throughout the city.[2]

It is reasonable to recognize that Washington's telegraph security from the Morse demonstrations of 1844 up to the outbreak of the Civil War was not generally robust. The confidentiality of telegraphic messages was mainly assumed rather than tested. The flow of the message along the telegraph networks from sender to receiver was known to be relayed by individual telegraph operators who therefore had access to the contents of the message, but the discretion of said telegraphers was rarely questioned. The ability to

THE FIRST TELEGRAPHIC MESSAGE FROM CALIFORNIA.

The establishment of the Transcontinental Telegraph (Library of Congress Prints and Photographs Online Catalog, reproduction number LC-USZ62–116943)

tap into a telegraph line and surreptitiously listen in on telegraphic traffic was somewhat unknown by customers. The possibility of receiving a false or deceptive telegram as a tactic of disinformation, whether it be inaccurate information or false attribution of the sender, was largely unrecognized. A related tactic was blocking, refusing, delaying, or neglecting the delivery of a telegram. Security procedures for receipt, storage, or disposal of the telegram (once transcribed or printed on paper) or for the written messages of the sender and receiver before and after the messages were telegraphed seem to have been largely nonexistent. A plethora of telegraph companies existed, and which companies might transmit part or all of a telegram from sender to recipient was often unclear. Whether anyone had significant security concerns about such matters remains unknown, for scant evidence exists to suggest telegraph users at that time saw these as security concerns. Users knew that individual companies, lines, and circuits could fail, that new commercial telegraph start-ups and networks could be lucrative investments, and that telegraph companies also could be unreliable as business ventures, for all these outcomes happened regularly. Basically nothing in the evidentiary record, however, exists to indicate these infrastructural instabilities also were widely seen and effectively acted upon as telegraph security concerns. When it came to telegraph security before the Civil War, most people, including those serving in government, mainly appear to have been uninformed, uneducated, and unwitting.

Disruption, Sabotage, Severed Lines, and Early Censorship

In April 1861, with the secession of southern states underway and in the immediate aftermath of the shelling of Fort Sumter, Washington-area militia were organized as an initial security force for the city and the District. The paramount security concern for the local militia and the Lincoln administration was the potential for the city to be attacked and taken by Southern sympathizers, a secessionist militia, or a Confederate army. Measures to address this concern included protecting roads, railways, rivers, and telegraph lines. Telegraphic connections between Washington and Baltimore (and by extension the Northeast) were briefly severed on April 15, 1861.[3] Union militia seized the telegraph office on Pennsylvania Avenue the evening of April 18.[4] Service and signal flow into and out of Washington remained sporadic for more than a week as telegraph lines were repaired. Many lines into Washington had been sabotaged and, on the evening of April 21, all telegraphic connectivity northward was reportedly lost. Confederate secessionists and sympathizers seized the main telegraph station in Baltimore.[5] Any sporadic telegraph traffic that could be sent north from Washington over the next few days other than official government messages was subject to government surveillance and censorship.[6] On April 25, a team of telegraph repairmen riding into Washington on a train with the Seventh New York regiment disembarked to repair downed lines. Andrew Carnegie, a member of the repair crew, had his face cut by a booby trap as he repaired a downed and sabotaged telegraph line.[7] Elmer Ellsworth, generally considered to be the first Union Army casualty of the Civil War, was killed in Alexandria on May 24 after cutting down a Confederate flag flying at the Marshall House, a local hotel. Ellsworth had been distracted by the flag and stopped his original mission: he was on his way with troops to seize the local telegraph office.[8] A telegraph line between Washington and Richmond was severed on April 21 by telegraph officials.[9] On June 1, individuals in Washington associated with the American Telegraph Company met on a bridge over the Potomac River and cut the company's telegraph line connecting to the south.[10]

Telegraph security for Washington in April 1861 was a geographical paradox. In many cases, the further away a telegraph line or station from Washington, the more secure the line or station, at least in some compass directions. Telegraph security for Washington to the South and the Confederacy was rendered immediately nonexistent. As early as January 1861, telegrams sent by southern sympathizers about weapons purchases and government decisions drew suspicion and led to some informal censorship by telegraph companies.[11] In May and June 1861 Arkansas residents and irregular militia along with Confederate troops attacked and destroyed commercial telegraph lines operated in Arkansas by Union sympathizers.[12] To the Northeast and the Eastern Seaboard, telegraph security was endangered. The oldest regularly operating telegraph route in the nation between Washington and Baltimore, inaugurated by Samuel Morse in 1844, was routinely sabotaged in the first few months of the Civil War. Confederate sympathizers attacked Washington–Baltimore telegraph connectivity with impunity, on many occasions with the tacit or even overt support of Maryland political leaders sympathetic to the Confederacy. Baltimore was a key hub for telegraph traffic between Washington and New York City, Boston, Philadelphia, and much of Pennsylvania, and by extension New England and the eastern cities of the Great Lakes. Nearby alternative telegraph lines and routes in and around the District of Columbia, Maryland, and northern Virginia that could bypass the Washington–Baltimore route also were attacked, tapped, damaged, sabotaged, filled with misleading information, and overall suspect. The situation in Baltimore and the Maryland attacks, sabotage, and subterfuge against telegraph security for Washington fostered imminent and growing communication isolation for the Union government. Secure command and reliable signal flow of many nearby and vital telegraph lines, stations, and networks quickly disappeared.[13]

In other directions from Washington, telegraph security was, by comparison, stronger and more greatly assured. Security for signal flow improved when one headed, or exchanged telegrams, between Washington toward the north and west, particularly to the northwest. Some telegraph corporations with most of their lines in southern states worked closely with the Confederacy. As Robert Luther Thompson noted, certain telegraph companies were more advantageously situated to respond to the Civil War in their locations and in the directions in which their particular lines ran. "Those extending from north to south across the Mason-Dixon line and the Ohio River were soon cut in two" but those running from east to west were "situated to link the loyal states with one another" and as a result "found the war years a period of golden harvest and growth."[14] Beyond the contested security conditions for the telegraph between and around Washington and Baltimore, as well as some other contested areas near the District of Columbia, security and signal flow problems somewhat abated. Although sabotage was always possible anywhere, the Great Lakes region had few if any such cases. Northward along telegraph lines and networks to Michigan, Wisconsin, and Minnesota was secure from war-related acts of subterfuge and sabotage. Southern areas of Ohio, Indiana, and Illinois raised some concerns for telegraph security.[15] Of course, telegraph lines and routes everywhere found instances calling for routine maintenance and suffered damaging conditions brought about from time to time by human error or environmental and climatological factors such as storms. Regarding the immediate security of the Transcontinental Telegraph, the original Missouri portion of the route was sabotaged by secessionists and was no longer secure. The governor of Missouri issued orders to tear down Transcontinental Telegraph lines within the state and thereby disrupt Union military communication.[16] The construction of a new northern route across Iowa in 1862 soon safeguarded the Transcontinental Telegraph from Chicago to Omaha.[17] From Omaha westward was another matter. West of Salt Lake City and onward to California and near San Francisco, war security remained generally reliable for the Transcontinental Telegraph, although there were occasional questions of Union loyalty about some inhabitants of the San Francisco Bay area. The security challenge for signal flow, state secrets, and secure command that remained even after the inauguration of the newer northern route of the Transcontinental Telegraph was the expanse west from Omaha along the Platte River, across the Great Plains into the Rocky Mountains and on to Salt Lake City. This portion of the Transcontinental Telegraph route would pose an increasingly complex set of security problems throughout and beyond the Civil War.

Albert Myer Takes Command

The rise of Civil War hostilities also brought orders to recall to Washington an army officer from the Department of New Mexico, where he had been engaged in applying his theories on flag, lantern, and field signaling as part of the Navajo campaign: Major Albert Myer.[18] After first stopping at Fort Monroe to establish a Signal Corps training camp, Myer was headquartered in Washington by August 1861 serving as both the signal officer of the Union Army (by 1862, chief signal officer of the Army) and as chief signal officer of the Army of the Potomac. Myer soon established a Washington-area signal training camp in Georgetown at Red Hill.[19] At that time, military signaling, aside from that conveyed by couriers, was largely accomplished by observation posts communicating with flags (by day) and torches or lanterns (by night) with interest in, but little development of, military telegraph applications and uses. Signal communication with flags and torches worked well under certain topographical conditions, such as the coordinated command of a chain of hilltops and high elevations. In June 1861, however, Myer informed Secretary of War Simon Cameron that other topographical conditions (such as valleys between hills) might necessitate alternative signaling such as by telegraph. Cameron authorized ten thousand dollars for investigation of military telegraphy but little, if any, military momentum toward telegraphy immediately ensued.[20] The expanding system of defensive forts in Washington, many situated on hilltops and other high elevations, did communicate by flag and torch. The Washington forts, and the Army of the Potomac, planned for telegraphic connectivity with telegraph trains (field units) and a rudimentary message system that "can be worked by any officer or intelligent soldier" in the field.[21] Myer visited the Red Hill camp nearly every day, usually in the evenings. In addition to the Red Hill site, Myer established Signal Corps offices in the District at 1905 F Street. (This area is now part of the campus of George Washington University.)

Myer intended to bring the telegraph into the Signal Corps and took several steps to do so early in the conflict. No clear plans or strategy for the telegraph and military activities, particularly in an internal rebellion, had emerged. For the army and the federal government during the 1846–48 war with Mexico, the telegraph was not a military asset, nor was it a means of communication linking the federal government with the battlefront.[22] American telegraphy had not extended to the battlefields of the war with Mexico. Military communication between the Mexican battlefields and Washington had circulated by military messenger and postal systems, carried on overland and maritime routes. Some newspaper accounts were "reported by telegraph" from the battle areas. In these cases, newspaper reporters used rapid overland travel (such as horse relay systems akin to the Pony Express) to reach the nearest operating telegraph station for subsequent relay along the nascent telegraph networks to the East Coast, a system that did provide more rapid reporting than did the military communication systems. Despite these moments, news flow from the battlefront was slow and newspaper editors "watched helplessly as the highly touted Morse telegraph inched through the south as war with Mexico raged on between 1846 and 1848."[23] President James Polk reportedly fulminated throughout the conflict about receiving war updates via newspapers more rapidly than he did from his military commanders. The telegraph was deployed by Great Britain in the 1850s as a military asset in the Crimean War, including the establishment of battlefield lines, equipment, and wagons for telegraph installations and operations. The British navy also shelled a Russian telegraph station located on an island near Helsinki, damaging the station and ruining the island's potato crop. What if anything the Union army learned from the British deployment of the telegraph as a military asset and a strategic target during the Crimean War was unclear. As Paul Scheips succinctly noted, "there was no military telegraph organization in the nation at the beginning of the (Civil) war."[24] For the defense and security of Washington, this lack of telegraph knowledge was not the exception but the norm. Essentially defenseless and without contingency plans when Fort Sumter was shelled, Washington was less understood from a defense perspective by American military engineers than were "the approaches to Paris and other European capitals."[25]

Before the Civil War, residents and visitors often saw Washington as a "city of magnificent distances" because of its relatively remote location from the populous areas of the nation. With war on its doorstep, the city now suddenly found itself "a city on the edge of danger, treason, deceit, political maneuver, uncertainty, and apprehension" with a military defense "more an illusion than a reality."[26] This was the case for everything, including the telegraph. The ensuing thirteen years from the 1848 Treaty of Guadalupe Hidalgo to the 1861 shelling of Fort Sumter were not marked by significant government study of or investment in telegraphy as a military or national security asset. Put simply, the escalation of tensions leading to the opening conflicts of the Civil War occurred at a time when the nation found itself more or less at a fossilized telegraphic status quo preserved from the war with Mexico. The telegraphic status quo was a military with no technical expertise, no lines or equipment, no battlefield tactics, and no doctrine, practices, or plans for telegraph security.

The United States Military Telegraph Company

For the first two years of the Civil War, what did not emerge were plans, policies, and doctrine to place the military telegraph with Albert Myer and the Signal Corps, but rather a distinct telegraph entity called the United States Military Telegraph (USMT) Company.[27] Instituted by Secretary of War Edwin Stanton as an intentional replacement for Signal Corps telegraphy, the USMT was created by Stanton, Thomas Scott of the Pennsylvania Railroad, and telegraph magnate Anson Stager. Composed of civilian telegraph operators rather than military personnel, the system coordinated and relayed Union military telegraphic communication. Stager and the USMT administratively reported to Quartermaster General Montgomery Meigs, but in terms of lived everyday experience Stager interacted directly with Stanton completely independently of field officers. Despite the 1862 nationalization of telegraph networks for the duration of hostilities, this arrangement largely allowed the operation of (Union) telegraph lines and networks to remain in private hands, albeit subject to the directions of Stager and Stanton.[28] Over the course of the conflict, as many as one thousand or more USMT operators saw wartime service, reporting to Stanton, Stager, and a small cadre of USMT officers. Virtually no USMT field operator had a military commission and instead nearly all served in "the anomalous status of quartermaster's employees, often under conditions of personal danger, and with no definite official standing" in the field. "It was simply a civilian bureau attached to the Quartermaster's Department, in which a few of its favored members received commissions."[29] Most of the telegraph operators faced the risks of conflict without the benefit of commission or rank as a member of the Union Army. The favored few with Union Army commissions as USMT officers held an "anomalous status" of difference in comparison with the telegraphers in the so-called civilian bureau. This difference was in holding the appointment and rank of a Union Army officer.[30] Nearly all USMT officers before, during, or after the war were associated with Western Union. The USMT represented one manifestation of the growing complexities and interconnections of the telegraph industry with the military branches of the federal government, complexities in growth and gestation from the onset of the Civil War. The complexities are discussed in further detail in chapter 5, particularly on the topic of USMT officers retaining military rank after the Civil War and overseeing the attempted construction of telegraph and cable lines in British Columbia, Siberia, the Bering Strait, and Alaska (Russian America) from 1865 to 1867.

Although it could be said that this treatment of the telegraph industry by Stanton had some common elements with the treatment of the railroad industry and the creation of the United States Military Railroads, there was a huge difference in leadership structure. The United States Military Railroads, while ultimately controlled by Stanton, was under the command of Herman Haupt, a railroad executive and bridge construction expert who became a Union Army officer in charge of military railroads for a portion of the war.[31] The USMT, on the other hand, was under the command of Anson Stager. Stager received, as did several other USMT officers, a military appointment from Stanton, and he held the rank of colonel. He did not have, however, anything at all resembling the respect and camaraderie Haupt and his staff enjoyed with Union Army officers, nor is it clear whether Stager tried to cultivate one. Haupt often was in Washington and occasionally met directly with President Lincoln. Stager spent most of his time during the conflict in Cleveland, not Washington. Major Thomas Eckert was the USMT officer of highest rank usually serving regularly in the nation's capital.[32] USMT operators and officers did not report to field commanders but only to a small internal command staff and on up to Eckert, Stager, and Stanton. Stanton and the USMT command staff did not want telegraph operators subject to orders from any officers in the field, including the command staff and generals of the various Union armies. Regular Union Army soldiers did on occasion look upon the field telegraph as an oddity or enemy device and sometimes cut a piece of line as a souvenir to send home. "Experienced stragglers" wary or weary of front-line duty typically became self-appointed experts in guarding telegraph lines for several miles to the rear. "At no time during any action was it difficult to get guards for the telegraph lines. There were thousands of volunteers." The telegraph operators in the field "were ordered to report to, and obey only, the corporation representatives who dominated the War Department," a situation which occasionally created "contempt or hostility" from Union Army officers.[33]

This structural tension was from time to time somewhat resolved through familiarity that developed between telegraphers and field officers.[34] General Ulysses Grant had many problems with the USMT operators, particularly in their occasional refusal of his direct orders to prioritize military messages on USMT lines, but Grant also admired many USMT field installers for their rapid and efficient work. He recalled in his wartime memoirs the routine daily details of setting up field telegraph lines: many teams of two men and a mule with reels of wire, wagons with telegraph equipment for armies, corps, divisions, regiments, and his own headquarters, other wagons with lightweight poles for stringing wire above ground. All were quickly joined together to make a "continuous wire in the rear of the whole army" linking together the army in the field and taking only "a few minutes longer time than it took a mule to walk the length of its coil."[35]

Yet the situation with Eckert, Stager, and Stanton in complete and isolated command and control of the USMT created many problematic moments. Roscoe Pound concluded that on many occasions the "result was an incredible degree of non-cooperation with army and department commanders which . . . rendered the military telegraph well-nigh useless and sometimes made it a downright menace." The USMT operator or officer assigned to a field command headquarters was, even in the thick of battle, an independent entity who answered to the Union Army commanding general "only when he (the telegraph operator) chose, and was quite indifferent to the exigencies of military operations." Western Union and the other commercial telegraph companies involved in the USMT apparatus demanded that their telegraphers continue to send, relay, and receive commercial telegrams not relevant to military matters while posted at battlefield locations.

The Civil War was an existential crisis, but from the perspective of commercial telegraph corporations it was not to be allowed to become an impediment to the unbroken private ownership and utility of the telegraph in American commerce, business, journalism, and finance. Many USMT operators augmented their income by sending private messages from the field, including military information that might be parlayed into stock market and trade tips, such as the probability of incipient price fluctuations in the cotton market resulting from recent military developments.[36] Such USMT operators on the battlefront with Union Army forces cultivated a "strong temptation" and "habitual tendency" to "give precedence, even in military emergencies, to business and press dispatches." This included the seemingly benign, but often troublesome, practice of allowing military telegrams to enter the same queue as business and press telegrams without any markings, orders, or instructions for military priority. This looked a little bit like net neutrality in the middle of the Civil War, with no structured priority to first send and receive military telegrams. Thus, a military telegram would routinely wait its turn in the message pile while a USMT operator first handled other, nonmilitary, messages. This meant that that "private, paying messages were being given the right of way." In the daily operations of the USMT system, business practices could and often did trump military priority and national security.[37]

USMT and Corporate Security

These USMT habits and practices were partly in keeping with certain attitudes of telegraph corporations toward the war and telegraphy. In summer 1861, plans for telegraph systems and technical expertise within the Signal Corps were met with disapproval by telegraph corporations. They argued that telegraph systems and telegraph capacity within and operated by the military, even in wartime, ran counter to the known practices of American commercial telegraphy. Ownership and operation of American telegraph lines, equipment, networks, expertise, and systems must always remain in private hands. "Corporations interested in the telegraph as ordinarily used were opposed to the idea of allowing it, or permitting its management in the least degree from under their immediate control."[38] Albert Myer reported an early resistance in 1861 to the growth of Signal Corps telegraphy, meaning Signal Corps progress and plans for the telegraph were "thwarted, to some extent, by the action of persons who seemed to greatly desire that all the duties of electric telegraphy should be in the hands of civilians."[39] Some criticized Myer for his efforts "to monopolize army telegraphing" and for proposing military plans where "even commercial telegraphing was threatened." From this perspective, all would have gone perfectly well for the telegraph and the Civil War if Myer "had been content to leave telegraphing by wire to electricians" rather than bringing the telegraph into the hands of Signal Corps officers and troops.[40] David Hochfelder summarized this arrangement as "an uneasy hybrid, a telegraph system that served the military but was not part of it."[41]

These conditions and examples point to the pursuit of telegraph security via the USMT as a secure entity only unto itself, an entity without regard to Union commanders and armies needing direct connectivity to the President and expected to ultimately be under his direct communicative command. The ultimate safeguard of national security for the telegraph, even in a condition of existential crisis, was total and complete private ownership, operations, strategy, and decisions about military telegraphy. Anson Stager called this attitude a valuable result attributable to the fact that the telegraph companies "evinced a most liberal and loyal spirit" that guaranteed the government had access to "all the facilities" possessed by the telegraph companies.[42] This was military telegraph security devoid of any secure command beyond the private security of its corporate self.[43] A vision isolating the president from direct telegraphic connectivity with the Union Army, this philosophy and practice perhaps skirted with unconstitutionality. The telegraph was something that formed an unusual bond between the military and the corporate from the first moments of the Civil War. This was the first U.S. example of the emergence of the unusual bond between the military and the corporate over electronic communication networks, an unusual bond with a trajectory that continues to the present day. This unusual bond was at the time, and remains today, an ambiguous relationship with its own ever-evolving internal discourse and closed conditions of security and secrecy.

The Civil War and the USMT represented a transition for American electronic communication corporations and the military, a transition during which the military became more corporate and the corporate became more military. The transformative nature of this unusual bond first encountered with the telegraph and Civil War security continued beyond the Civil War and continues today. Some similarities between the telegraph and the railroad in this regard are discernible. Both were new technologies nationalized by the Union during the Civil War and exhibiting military-corporate cooperation during the conflict and in its immediate aftermath. The everyday practices and experiences that characterized the cooperation in the two cases proved markedly different. One of the most striking differences was the one between Haupt and Myer, particularly in their relationships with Stanton. Haupt earned Stanton's support (if grudging) and Haupt had the respect and collegiality of Union officers and commanders. Myer earned Stanton's enmity, yet many Union officers found Myer's signaling and communication systems effective.

Edwin Stanton and the Telegraph

Myer and the Signal Corps did have early difficulties with the corps's early telegraph systems and techniques at the battlefront, having first chosen to deploy a telegraph system with a range of only a few miles and what proved to be other shortcomings during operation.[44] Even if the Signal Corps had been successful with its first telegraph attempts, one wonders whether Stanton would have tolerated the Union Army basically running—through one of its own units—its own telegraph equipment and networks and training its own telegraph personnel. Stanton exerted complete control and mastery over the Union telegraph networks during the war, even prohibiting the installation of a telegraph station in the White House. Albert Johnson, confidential secretary to Stanton, recalled that Stanton's second act upon joining the Lincoln administration (after immediately ordering officers enjoying leisure time in Washington to return to their troops) was "creating the Central Telegraph Office in rooms adjoining his own, and directing that all war telegrams should go through this office with copies furnished to him." Soon after, Stanton ordered the telegraph station removed from Commander of the Army of the Potomac George McClellan's headquarters. Control of telegraph security continued to be firmly constructed and then held by Stanton throughout and after the Civil War. When battlefield hostilities ended and during his 1866 confrontations with President Andrew Johnson in the aftermath of the Lincoln assassination, Stanton took the macabre precaution of secretly hiding his voluminous copies of wartime telegrams in a vault at Ford's Theatre. Perhaps this was, in a nineteenth-century way, akin to a twenty-first-century government official storing e-mail and other government-related electronic data on a private rather than government server.[45] Stanton was—literally—the embodiment of telegraph security for Washington and the Union during the war. His control of the telegraph not only included military operations and tactics but also meant control and influence over the flow of telegraphic data concerning commerce, finance, governance, transportation, and news.[46]

After the conflict and before the return of telegraph lines to private owners (the return to private ownership ran from about August 1865 to June 1866), Anson Stager coordinated significant repairs to telegraph lines used in the USMT system. The USMT rebuilt lines in the former Confederacy and oversaw the construction of 5,256 miles of new lines in 1865–66, many miles of which were what were called the field telegraph lines. Stager's efforts were federally funded and the lines, once rebuilt, reverted to private ownership and operation. In some cases, line repair had already begun prior to the cessation of hostilities.[47] USMT personnel continued to operate many lines in the former Confederacy until December 1865. With the final repairs and the last round of construction of new lines, Stager finally closed the budget on the USMT in November 1866, kept his military appointment until 1868, and then moved to Chicago to serve as president of Western Electric.[48] By the end of the war, Western Union had absorbed many of its commercial competitors and by February 1866 was the largest telegraph corporation in the nation with more than seventy-five thousand miles of telegraph lines. Many more miles of lines would be added by Western Union in the coming two decades.[49]

The USMT budget closing of November 1866 was more or less coincident with the successful establishment of the Atlantic cable in September 1866. This had implications for Western Union and USMT activities in Russian Alaska at that time. As discussed later in this book, some USMT officers overseeing the 1865–67 Western Union Telegraph Expedition retained their military commissions and ranks and remained in uniform after the Civil War was over. They served in uniform on the ground in British Columbia, Russian Alaska, and Siberia during and after the Civil War, and after Stager officially closed the books on the USMT. The unusual bond carried over into a preoccupation with Alaska.

The control by Stanton over all aspects of telegraphy for the Union during the war included his refusal to establish a telegraph office in the White House. The president was first telegraphically isolated at the outbreak of the Civil War by the sabotage or capture of local telegraph lines and the lack of access to nationwide telegraph connectivity and secure signal flow that Lincoln and all Washington experienced in April and May 1861. Incredibly, Lincoln was then telegraphically isolated again, this time by his own secretary of war. The president's isolation from electronic communication networks, seemingly unthinkable today whether in peace or in war, denied him direct and independent access via electronic communication networks to Union military forces and their command staff, to state and territorial governors, to business and finance communities, and to the press—indeed, the whole of the Union. Access to the telegraph for the president of the United States during the Civil War was constricted, policed, and monitored by the secretary of war.

Like everyone else (save the commercial telegraph companies) in Washington during the war, Abraham Lincoln could access the telegraph only through the system Stanton controlled. Despite this astounding condition for the president of the United States, Lincoln made extensive use of the telegraph while president, most often by walking over from the White House to visit the War Department Telegraph Room. This was a near-daily routine for Lincoln, where he also perused dispatches from battle areas and conversed with telegraph operators. Stanton had at one point tried but failed to prevent Lincoln from even seeing telegrams sent to the War Department from the various battlefronts. Stanton insisted that originals of all telegrams from Union Army generals or officers in the field be sent by USMT operators to the War Department. Stanton at the same time kept copies for himself of any telegrams sent by President Lincoln.[50]

Blocked by Stanton from having something akin to a Situation Room directly in the White House, Lincoln made the War Department telegraph office his own de facto Situation Room for electronic communication. Even this form of access was remarkable, because it was not private and confidential access to the telegraph for the president of the United States. Rather, it was monitored access under a form of surveillance or supervision. Many anecdotes and stories confirm that Lincoln and many of the War Department telegraphers enjoyed warm, cordial, and supportive relationships. Those relationships did not, however, militate against or atone for the fact that Lincoln's telegraph access was not private, but rather semipublic, routinely supervised and in the presence of others at the War Department Telegraph Office. Perhaps this was a bit like contemporary users of laptop computers, Blackberries, smartphones, Bluetooths, and other technologies of mobile social media, working on the go at airports, hotels, and cafés. Today these users are engaging in electronic communication while observable in semipublic or public settings and not in securely contained private physical locations and conditions: actions now associated with a wide range of people, but to this day not routinely associated with the president of the United States of America. To look back from our contemporary world and see this institution and enforcement of monitoring, surveillance, and conscribed affairs for the president of the United States and the telegraph in the Civil War still stuns and shocks.

The USMT and the Trans-Mississippi West

Stanton, Stager, and Eckert also succeeded in isolating Albert Myer and the Signal Corps from independence and autonomy in the USMT. A series of actions and decisions by Stanton, Stager, Eckert, and the USMT command structure in 1863 and 1864 removed all telegraph activity from Signal Corps command, stripped the Signal Corps of its telegraph equipment, transferred that equipment to the USMT, and halted Signal Corps telegraphy in the field.[51] Myer was removed as chief signal officer. He eventually found himself going back and forth across the Mississippi River arranging signal communication for Union armies in the Gulf, Western Mississippi, and Arkansas.[52] In 1864 he was under the command of General Edward Canby leading the Army of West Mississippi. Myer had first been under the command of Canby in New Mexico Territory in 1859, where Canby led a campaign against the Navajo and Myer tested his wig-wag signal communication theories and systems. In 1864, in addition to providing Canby with signal support, Myer added the task of prisoner and deserter interrogation to Signal Corps duties. This included individual interrogations and the collation and comparison of information from individuals for further analysis across deserter and prisoner samples.[53] Surveillance and interrogation practices begun by Myer and the Signal Corps during the Civil War would return to the Signal Corps during the Railroad Strikes of 1877 and for the indigenous peoples of the Southwest, as discussed later in this book. The USMT also operated across the Mississippi River in Arkansas in 1864, although the USMT found it impossible to keep the telegraph line to Fort Smith in working order for most of the year because of guerrilla activities. General William Rosecrans enjoyed more success in Missouri that same year with secure USMT lines emanating from his headquarters in St. Louis, including a line to Fort Leavenworth and Fort Scott in Kansas. This prompted General Samuel Curtis to proclaim to Stanton that he needed "telegraphic communication" and the "Arkansas River must be lined with posts (telegraph poles), and department commanders must have connection with them everywhere" but the USMT's call for more poles, lines, and telegraphic communication on the Arkansas River went unheeded by Stanton and Stager.[54]

The relative inattentiveness of the USMT to Curtis, the Arkansas River, and other Trans-Mississippi regions had implications for the Transcontinental Telegraph, particularly as concerned information about gold mining in California and in Colorado Territory. Revenue from gold mining was an important financial factor for the Union in waging the Civil War. The USMT operated mainly east of the Mississippi River. The USMT did, however, build and operate field telegraph lines in 1862 and 1863 (generally speaking, during the Civil War, field telegraph lines refer to temporary telegraph line installations coordinated with troop movements and battlefields) in Missouri, Kansas, Arkansas, and Indian Territory.[55] Despite crossing the Mississippi River into Texas, Arkansas, Missouri, and portions of Kansas, the reach of the USMT across the Trans-Mississippi West turned out to have geographical limitations. In retrospect, the limitations appear to be somewhat related to the perceived potential, or lack thereof, for profitable commercial telegraph enterprises west of the Mississippi River in the aftermath of the Civil War. The potential for immediate commercial profitability in the Confederacy east of the Mississippi was obvious—beyond the Mississippi, not so much. The Arkansas River area presumably did not yet seem ripe for postwar profit by commercial telegraph companies. Anson Stager had assured commercial telegraph companies as early as 1861 that the USMT and the federal government would not own and operate government lines after the Civil War and that any such lines built during the war would be available for purchase and subsequent operation by private companies after the conflict was over.[56] The one commercial telegraph line in western North America that was immensely profitable was the Transcontinental Telegraph. In 1863 the value of stock owned by Western Union in various telegraph lines in Kansas, Missouri, and Arkansas totaled to $82,862.00, but the stock held that same year by Western Union in lines that were part of the Transcontinental Telegraph was valued at $2,504,964.00.[57] The geographical limitations of the USMT were evident in the problematic security of the Transcontinental Telegraph.

Generally secure from Chicago westward beyond Omaha to Fort Kearny in Nebraska Territory and also from Sacramento eastward to Salt Lake City, security for the Transcontinental Telegraph between Salt Lake City and Fort Kearny remained indeterminate, contested, and fraught. This middle stretch of the Transcontinental Telegraph through the Great Plains included more than twenty main telegraph stations, additional temporary or branch stations, hundreds of miles of line, and thousands of telegraph poles and insulators. The Transcontinental Telegraph was electrified by storage batteries, and that required a telegraph relay station every fifty to seventy-five miles along the line.[58] The number of telegraph stations along the Transcontinental Telegraph, particularly in this middle stretch, varied over time as forts, posts, and communities were built or abandoned. As a result, the line is better understood not so much as a fixed and permanent entity of established stations, but rather as a telegraph conduit that many people and locations all along the

Deer Creek Station (William Henry Jackson Collection, Scotts Bluff National Monument, National Park Service, catalog number SCBL_767)

route throughout the 1860s used for temporary or semipermanent access to telegraph services and communication. In this sense, the Transcontinental Telegraph was not like a contemporary multilane limited-access highway with demarcated entrances and exits. It was more like a trail or a river that had major access points but could be, and often was, accessed at any time and at any point along the line.

Among the major telegraph stations in this stretch were Fort Bridger, Fort Laramie, and Julesburg, the latter the only station in Colorado Territory and the only telegraphic link to the Transcontinental Telegraph for Denver and the gold fields. Denver had initially balked at local fundraising to support a Transcontinental Telegraph station, and the city did not have a telegraph branch line to Julesburg until October 1863, a branch line then built by Edward Creighton.[59] Even after that line opened, problems continued. The Denver–Julesburg branch line was not particularly reliable and had its share of malfunctions and failures to accurately relay messages. Before the Denver–Julesburg line, telegrams in and out of Denver traveled by overland roads back and forth to Julesburg, where the written messages were then transmitted and received by the Julesburg station. Telegrams received were then written down and transported back to Denver. The October 1863 Denver branch line alleviated, but did not fully eliminate, the transition from written message out of (or back to) Denver for a telegram on the Transcontinental Telegraph sent or received at Julesburg.

Poles, lines, wagons, equipment, and stations could be targets for damage or sabotage. In a land with scarce timber, a telegraph pole was a tempting target for someone wanting a cooking fire. Even in and around Fort Kearney the telegraph poles were on occasion targeted, not for sabotage but for shooting practice. One local resident was skilled at hitting up to a dozen telegraph poles with bullets from two revolvers as he rode a horse and galloped along the line.[60] Residents and ranchers along much of the Transcontinental Telegraph, particularly between Fort Kearney and Salt Lake City, tested and maintained the line. Test signals were sent over the line three times a day. Settlers and ranchers along the line used a device made of two brass plates that, when a wet finger was swiped along the plates, would send a small shock ensuring the signal was flowing and the line operating. If no signal was detected, "he was instructed to wait for fifteen minutes; then, if there were still no signals, he was to get his horse ready." Loss of signal detection longer than 30 minutes necessitated riding to the nearest telegraph station, looking for any broken lines.[61] The Transcontinental Telegraph and its branch lines also aided in surveillance and capture of Union Army deserters, who often simply followed a telegraph line as they drifted away from their post. Operators witnessing a likely deserter telegraphed a frontier military post to inform the post and await instructions. A telegraph pole proved handy when the need arose to tie up a town drunk or a local troublemaker for a day or so.[62]

Native American Resistance against the Transcontinental Telegraph

Native Americans knew of the telegraph and its construction. Many accounts of the era tell of early encounters between Native Americans and telegraph construction crews and operators. Some accounts offer a range of stories from cooperation to antagonism, and other accounts have less to say about Native Americans. Several accounts relate meetings, discussions, and the presentation of gifts to indigenous peoples by some of the telegraph financiers, builders, and operators.[63] One commonly repeated account from the early 1860s, with some variation, detailed Native Americans cutting down telegraph wire, riding off with the wire, encountering an electrical storm, and receiving an electric shock from the wire. Some accounts suggest intentional trickery on the part of some telegraph builders, who encouraged one or more Native Americans to hold a bare wire, through which the telegraph builder then delivered an electric shock. In his 1947 study of the American telegraph industry, Robert Luther Thompson breezily commented that west of Salt Lake City "there was never any hostility between the construction crews and the red men" but between Salt Lake City and Omaha "unfortunately, there were some exceptions to this happy relationship."[64] These accounts written by newspaper reporters, telegraph builders, white settlers, and military personnel in diaries, letters, and journals typically narrate a tale of dawning Native American self-awareness that the electrical shock from the telegraph wire was "punishment" and the telegraph was "bad medicine."[65] These accounts were then retold, often verbatim, by many newspaper reporters, adventure writers, military personnel, and some scholars. In repeating these tales, Eugene Ware claimed that these stories of the telegraph delivering punishment and bad medicine through electric shocks "circulated with great rapidity among the Indians, and none of them could ever afterwards be found who would tamper with the wire." He added Native Americans would still "cut down a pole and use the wood for cooking." Ware and many others who told or recounted these stories usually did not add that the hewing and burning of a telegraph pole for a cooking fire was more commonly done by white settlers and trail travelers. When poles were damaged or burned and needed replacement, the old pole stumps were saved by the replacement crews for firewood. Wood was always in scarce supply.[66]

Escalating tensions between Native Americans and white settlers in Colorado Territory during 1864 led to an increasing series of reciprocal attacks and reprisals by indigenous people and white settlers. John Evans, governor of Colorado Territory, telegraphed Stanton on more than one occasion to request that the secretary of war assign Union troops to Colorado Territory. Evans did not at first take the initiative to raise local troops and form regiments from territorial volunteers on his own, perhaps because his predecessor, William Gilpin, had been removed from the territorial governorship for independently financing a territorial volunteer military regiment without prior approval from Stanton. Evans also corresponded by both mail and telegram with the commander of the Department of Kansas, General Samuel Curtis, about territorial military protection from Native Americans.[67] Stanton did not support the request for regular army troops from Evans, and instead of Union troops, local militia and volunteer regiments eventually became the norm. In spring 1864 crop failures, scarce game, and the lack of promised government supplies led some Native Americans to raid Colorado Territory farms, actions met by retaliatory attacks led by John Chivington.[68] In August 1864 the military command in Washington authorized Colorado Territory to recruit and form a volunteer regiment for a period of one hundred days.[69]

Through August of that year, Native Americans had not sabotaged or damaged the Transcontinental Telegraph beyond the occasional burning of a pole. Attacks on telegraph operators from settlers and other adventurers did happen from time to time. Hervey Johnson wrote home in October of 1864 about an "old mountaineer" who "attacked a telegraph operator with a hatchet" over a perceived breach of promise.[70] In August 1864 Native Americans "cut the wire three times between Valley Station and Cottonwood Springs" in what was considered the first deliberate and militant attack by indigenous peoples on the Transcontinental Telegraph and its branch lines.[71] All month long, relationships between Native Americans and white settlers in Colorado Territory continued to deteriorate and tensions increased. In October, Overland Stage owner Ben Holladay wrote Stanton asking for more protection for his stage lines from possible raids and attacks by Native Americans and recommended assigning Patrick Edward Connor of the military district of Utah to quell Colorado Territory native uprisings, a request quickly approved by Army Chief of Staff General Henry Halleck. This series of events increased the flow of letters and telegrams on the already-bellicose militant talk of the Colorado territorial government and militia, particularly inciting the tongue of Chivington, commander of the Third Colorado Cavalry.[72] From mid-August to mid-September 1864, mail headed east from Denver could not be safely sent via Julesburg and instead traveled west to San Francisco, whence it then circled back east by way of Central America.[73]

From Sand Creek to Julesburg

All came to a head in the gruesome and murderous atrocity of the Sand Creek Massacre that Chivington led, mercilessly committed against Native Americans on November 29, 1864. Some Coloradans, along with many in eastern North America, looked upon Sand Creek with disdain. Many other Coloradans, particularly the citizens of Denver, welcomed Sand Creek, calling it an example of what must be done to indigenous peoples: extermination.[74] After the Sand Creek Massacre, "the Indians immediately put the Platte River under surveillance."[75] What came in the immediate aftermath of Sand Creek was a series of Native American reprisals across the Great Plains, including a long-running series of attacks on the Transcontinental Telegraph and various branch lines that soon led to "the systematic destruction of the Denver–Julesburg line and its telegraph offices."[76] Telegraph stations at Julesburg and Sterling were attacked on January 7, 1865, and the telegraph station at Junction Station was attacked on January 15. Wires were cut and hauled away and poles and stations were burned. Julesburg was attacked again in February, as was the telegraph station at Mud Springs, as well as every ranch and station within twenty miles from Julesburg.[77] This was retribution for Sand Creek with, as William Unrau noted, "the certain intention of destroying the telegraph and closing the overland routes." These attacks expanded and continued into the summer months, "disrupting telegraph service, discouraging emigration, and providing the beleaguered garrisons at St. Mary's, Three Crossings, Sweetwater, and Deer Creek with no lack of action."[78] During one attack near St. Mary's, troops took shelter in a nearby cave. One soldier brought along a length of wire and a telegraph key, ran a temporary connecting line, and from the shelter of the cave reported to Fort Bridger.[79] The mail and stage routes also were attacked. In the middle of January, overland mail from eastern North America to California was suspended, and some mail was returned to New York City and sent to San Francisco "via the isthmus route."[80] In February 1865, the territorial legislature of Nebraska Territory petitioned Congress to pay the expenses of the Nebraska Territory militia because the militia was protecting trails and telegraph lines against Native American attacks on those trails and lines. Cognizant that the "suppression of the slaveholder's rebellion is an act of Herculean proportions" the territorial legislature reminded Congress that the Nebraska militia helped ensure that "the millions of treasure annually exhumed from the great mining regions of the west" safely traveled "to the east over the great overland route."[81] In June 1865 one soldier wrote home that the "Overland Mail road from Julesburg to Green River in Utah is completely in possession of the Indians." A wide-ranging assault on the new transportation and communication networks of the Great Plains was in full force. "The attacks on Julesburg involved large numbers of Native Americans and were well organized and executed with military-like precision."[82] Great Plains warriors succeeded in deploying a nineteenth-century version of ongoing denial of service attacks on these networks.

By July 1865 some army officers had become reluctant to send out a telegraph repair patrol. One lieutenant thought small repair patrols resulted in "men butchered up needlessly" and the security of Great Plains telegraphy could be met only by significant troop reinforcements from the War Department. By October 1865, troops were pleading with their unit commanders, sometimes with success, to carry out actions and travel that "would avoid

Attacking Julesburg (Denver Public Library Digital Collections, call number X-9612)

places where they could be reached by telegraph."[83] Such actions and attitudes contradicted standing orders for all military personnel to "promptly repair the wire whenever required without waiting for orders to do so."[84] Great Plains warriors "split a great many telegraph poles to make travois in which to transport their dead and wounded."[85] With Civil War battlefield hostilities now months in the past, the telegraph lines and routes of the Great Plains had become the most dangerous territory on the continent. These telegraph lines were death traps. As a result, line repairs were often temporary, allowing wire to rest on the ground or be propped up with sticks to keep the line open while awaiting more substantial repairs and pole replacement. Throughout 1865 and into 1866, army units protecting telegraph lines on the Great Plains often found themselves poorly equipped with defective armaments and shells. Army horses were unaccustomed to grazing on prairie grasslands. Troops suffered delays in delivering military supplies, uniforms ill-adapted for the extremes of local climates, and subpar food and provisions.[86] In the summer of 1865, telegrams meant for transmission on the Transcontinental Telegraph were delayed as much as a month.[87] In addition to the gold fields of California and Colorado Territory, miners in Montana Territory in the 1860s found significant gold deposits, adding to the potential income of the Union government and also heightening the significance of the Transcontinental Telegraph for updated information about gold revenue flowing eastward from the Trans-Mississippi West.[88] All the while, plans were made and progress slowly gained on building a new transcontinental infrastructure: a railroad, completed in 1869.

What ensued in the aftermath of Sand Creek was the first sustained, long-term, ongoing attack by indigenous peoples against the means of electronic communication wielded by an occupying power on the North American frontier and perhaps anywhere in the world. This sustained series of attacks on the telegraph by Great Plains warriors was not like the British tactics against Russian telegraphy in the Crimean War, nor like the tactics of Southern sympathizers disrupting telegraphy near Washington in April and May 1861. In terms of attacking telegraph systems, it surpassed in magnitude and destruction the 1861–65 Civil War telegraph sabotage by Confederate sympathizers in Maryland and elsewhere on the Civil War battlefields in eastern North America. A cycle of telegraph repair and telegraph attack set in across the Great Plains. At a February 1865 encounter between troops and Native Americans near Julesburg, artillery gunners discovered they had lost the priming wire used to poke a hole in the flannel bags of powder loaded into their field pieces. Troops chopped down a nearby telegraph pole, cut

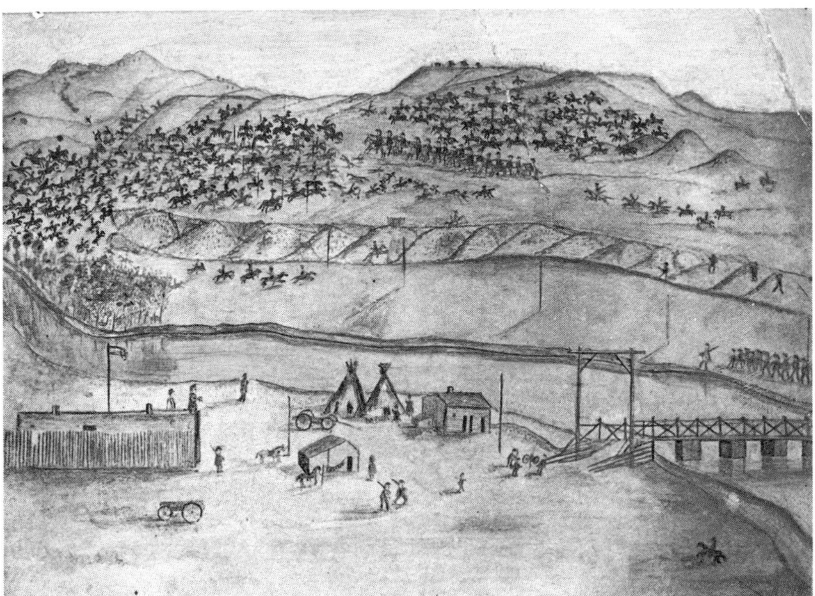

The Battle at Platte River Bridge in Wyoming (Collection of the Kansas Historical Society, item number 227202)

off a section of the line, spliced the line to restore signal flow, and used the telegraph wire as a priming-wire substitute. The evening of that battle, Native Americans built a huge bonfire out of telegraph poles.[89] The most severe damage took place on the telegraph line connecting Denver and Julesburg, but damaging attacks took place at many telegraph lines and stations. Repairs to the Denver–Julesburg line did not end Native American attacks on the telegraph: "On the contrary, it marked the beginning of a period during which the Indians systematically cut wires and burned telegraph poles as part of an ongoing guerrilla action against the government of the United States and its agents."[90]

Scorched-Earth Warfare on the Grassland: The Great Prairie Fire of January 1865

On January 27, 1865, "the sky was bright and clear, with a keen wind blowing from the northwest." Near Cottonwood Canyon in Nebraska Territory at sunset that day, General Robert Mitchell ordered the ignition of a coordinated prairie fire to drive away Native Americans. Despite all the backed-up traffic and delayed telegrams from California and in direct contradistinction to USMT practices against delaying commercial telegraph traffic to prioritize military messages, Mitchell "cleared the telegraph line early in the forenoon, and wired instructions up and down the (Platte) river." Mitchell ordered "that at sundown the prairies be simultaneously fired from Fort Kearney west to Denver. Instructions were sent to every ranch and post along the line." The fire was simultaneously lit just to the south of the Transcontinental Telegraph route and the Denver–Julesburg branch line. With a fire lit to the south and the telegraph lines protected by a prevailing northwest wind, the initial conflagration "fired for three hundred miles." Spread by horses pulling burning bales of hay and fed by the wind, the fires were at first "scattered beginnings; they were soon united, and they rolled as a vast confluent sheet of flames to the south." The army used a coordinated raging wildfire on the Great Plains as a weapon of mass destruction against indigenous peoples—all indigenous peoples, not just active indigenous warriors. The telegraph key ignited the simultaneous ignition over three hundred miles and transformed the Transcontinental Telegraph from an asset of military security into the catalyst for military destruction of a huge ecosystem, an action supposedly justified by the waging of war. "The fire swept the country clean; three days afterwards it was burning along the banks of the Arkansas River, over which river it passed in places and ran out down in the Panhandle of Texas."[91]

The prairie grassland that provided nutrition for horses and bison was incinerated, and a huge swath of the ecosystem central to the lives of Great Plains people was laid to waste.[92] The grassland scorched by the fire between the Platte River and Texas Panhandle became for a time generally uninhabitable for indigenous peoples. They could no longer "live in an area of territory which had been burned over and cleared of game; nor could the game stay during the winter in a country that was burned over."[93] It is likely that throughout 1865 and perhaps also in 1866 some bison herds headed north from Texas and, unable to fully graze on the burned-out prairie sections, chose an eastern bypass around the burn area. If so, they were no doubt slaughtered in great numbers by white hunters as they neared the central and eastern sections of Nebraska Territory. The Great Prairie Fire and its lingering impact on Great Plains wildlife may have helped fill the bellies of westward moving railroad construction crews with an abundance of bison meat. This was scorched-earth warfare deployed against Native American warriors, women, children, elderly, animals, communities, and possessions and deployed as well against the prairie ecosystem at the core of indigenous Great Plains culture and society. It was a nineteenth-century application of what Jacob Darwin Hamblin called arming Mother Nature.[94] It was also, in a way, a nineteenth-century version of the telegraph as the internet, but not a "Victorian" version of the internet.[95] When the Transcontinental Telegraph relayed the messages that triggered the Great Prairie Fire, the Transcontinental Telegraph presaged the

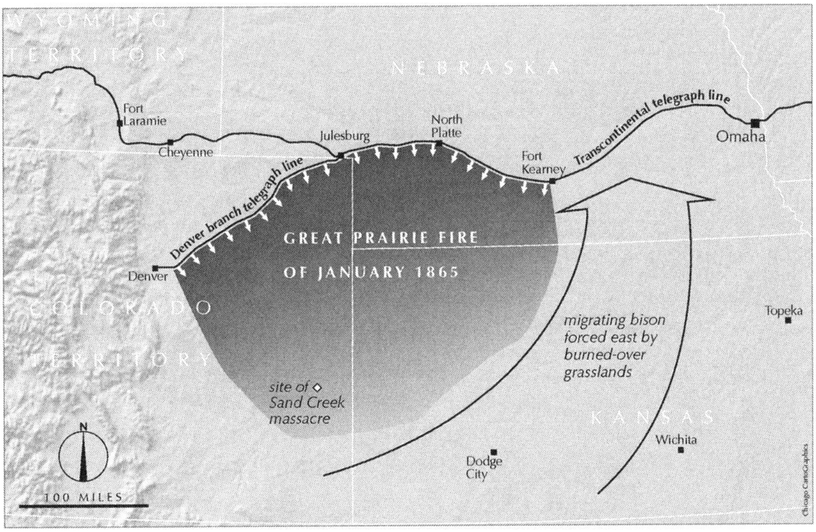

Map of the Great Prairie Fire of January 27, 1865 (map by Chicago CartoGraphics)

ultimate (but never issued) prime directive of ARPANET: relaying messages to deploy American weapons of mass destruction in a nuclear confrontation with the Union of Soviet Socialist Republics.[96]

Scorched-earth tactics had already taken place on the Civil War battlefields of eastern North America. Two well-known examples were the November–December 1864 march to the sea across Georgia from Atlanta to Savannah by General William Tecumseh Sherman and his troops, and the November–December 1864 burning of crops across portions of the Shenandoah Valley in Virginia by troops commanded by General Philip Sheridan. Both Sherman and Sheridan carried out wide-scale destruction of crops and farm animals. The march to the sea placed a total area of about nine thousand square miles of Georgia at risk of scorched-earth tactics, and the burning of the Shenandoah Valley placed about three hundred square miles of Virginia at risk. By comparison, an approximation of the area at risk from the Great Prairie Fire approached 40,000 square miles, an area equivalent to the entire state of Ohio.[97] Much of the scorched prairie was not yet inhabited by white settlers. The Santa Fe Trail ran through some of the burn area, although that trail was much less extensively used than other trails during the Civil War.[98]

Telegraph service on the Transcontinental Telegraph and branch lines between Fort Kearney and Salt Lake City remained sporadic and intermittent for nearly two years because of regular attacks and sabotage of those lines by Great Plains natives. Citizens in Denver were "in great anxiety" over the ongoing situation of downed lines and pleaded for service to be restored for their own protection and because "a great quantity of messages needed transmission. . . . No communications could be had with Laramie, Salt Lake, or the Pacific Coast, owing to the destruction of the telegraph line."[99] The attack on Julesburg in February 1865 created more telegraph havoc with buildings in "smoldering ruins" and downed telegraph poles and wire "twisted and tangled inextricably" scattered over a stretch of nearly fifty miles.[100] Military detachments became uneasy when any telegraph lines were out of commission. In some cases, word had been received by telegraph at a military detachment of an attack underway at a station or fortification elsewhere along a telegraph line, and then suddenly the line went dead, meaning "we just have to wait in suspense" for a line repair or news from other troops arriving at a fort or station.[101]

Indigenous Knowledge of the Telegraph

Whether and what Great Plains natives precisely and in detail knew of the telegrams between Evans, Curtis, and Stanton discussing requests for army troops and authorizing a temporary volunteer militia for Colorado Territory remains unknown. Also unknown is the precise detail of Native American knowledge and awareness of the growing correspondence of letters and telegrams sent via the Transcontinental Telegraph and branch lines, which were increasingly calling for military action against Native Americans. In their research relevant to this question, Thomas Jepsen and James Reid both suggested that Native American knowledge about telegrams and telegraph traffic was plausible, if not provable. Jepsen adverted to their knowledge in his research on the telegraph and Colorado Territory, which he based on speculative newspaper accounts from the era and his own inference. Reid recounted that early fears that Native Americans would attack the Transcontinental Telegraph proved unfounded "until the hostile tribes learned of its connection with army operations, when, for a time, trouble ensued." Scholars such as Jepsen and Reid deserve credit for introducing the plausibility of indigenous knowledge about the telegraph beyond the oft-told discriminatory and primitive tales of the telegraph bringing punishment and bad medicine.[102] A telegraph signal testing set from a rancher, a tap wire, or continuing conversations with local ranchers, farmers, trade agents, or Confederate sympathizers (who had arrived in Colorado Territory and the Great Plains during the Civil War in increasing numbers) may have yielded sufficient information about the content of telegrams between Denver, Fort Leavenworth, and Washington. Access to the Transcontinental Telegraph was on the one hand limited by the small number of telegraph stations and a growing, but not extensive or robust, number of locally connected branch telegraph lines. On the other hand, access was not secure beyond the builders, operators, and telegraph stations. Farmers, ranchers, construction crews, and others along the route with sufficient technical expertise and equipment could and did tie into telegraph lines and thereby could and did access the signal flow of the Transcontinental Telegraph and its various branch lines. Telegraph operators on the Transcontinental Telegraph and other lines routinely discussed telegrams and telegraph traffic at their stations with visitors to their stations or in the local communities. People visited telegraph stations and conversed. Telegraph stations and newspaper offices were places of public habitation, social settings, gossip, and business.[103] The practices, habits, and sociotechnical infrastructure around the signal flow of the Transcontinental Telegraph and its branch lines could not successfully secure state secrets. If telegraph access and information in Great Plains telegrams could even be called a secret, it was a completely open secret.

News and information received by the Transcontinental Telegraph was more often than not routinely discussed beyond the senders and recipients

Native Americans encountering the railroad, telegraph, and bison (Collection of the Kansas Historical Society, item number 220462)

of telegrams and was, of course, standard fodder for newspaper stories and editorials in Omaha, Fort Kearney, Salt Lake City, Denver, and many other cities and towns along the telegraph route.[104] The signal flow was not reliably secure. Neither were the messages on that signal flow. Even in ciphered telegrams, the names of the sender and receiver were not particularly secure. Even without individual names attached to messages, if one knew the location of a sender or receiver, it was not difficult to understand what it generally meant for telegrams to go back and forth between Denver and Washington, or Denver and Fort Leavenworth. Beyond the telegrams, the security measures taken with written messages that were then sent or received via telegraph remains unknown. There seems little reason to suspect such messages were routinely and carefully kept from others, properly disposed of, secured in a strongbox before and after transmission or reception, or fully kept secret from whoever carried the message to and from the telegraph station. The signal flow on the line was not particularly secure. The many written messages and additional paperwork associated with sending and receiving telegrams on the Transcontinental Telegraph and its branch lines, particularly the stretch between Fort Kearney and Salt Lake City, seem even less secure than the telegraph signal itself. Like wood, paper also was a scarce commodity on the

Great Plains, and the back of one message might be used for writing down something else. The secure command of the Transcontinental Telegraph was a mirage.

Put simply, the signal flow and telegrams on the Transcontinental Telegraph and branch lines of Colorado Territory were not securely guarded secrets. They were closer to a common topic of general conversation. In some ways, the Transcontinental Telegraph during the Civil War, especially the stretch between Fort Kearny and Salt Lake City, was akin to a contemporary internet service or application without user encryption, log-on procedures and passwords, secure electronic document storage and destruction, or any other form of network and data security. It is unimaginable today that a reasonable person with even a rudimentary understanding of the internet would knowingly use an unsecured internet service or application for confidential communication. The porous security of the Transcontinental Telegraph serves as one of countless historical reminders that secrecy and security with electronic information networks never was and is not today self-evident. It is instead a socially learned and constructed process that, to this day and into the foreseeable future, ever remains in contention and flux. This contention and flux in the security and secrecy of electronic information networks is not a new or recent condition of contemporary society. It is a long historical arc going back to the origins of the telegraph and subsequent electronic communication networks, and an arc seen in distinct and sharp relief from the perspective of the telegraph and western North America.

The Problem of Primitivizing Others

A lingering unvoiced assumption of a sort pervades many of the archival documents, newspaper accounts, and some research scholarship on the Transcontinental Telegraph, Sand Creek, Julesburg, and violent confrontations between indigenous peoples and white settlers, militias, territorial governments, and the army during this period. This assumption is that the indigenous peoples of the Great Plains and the territories encompassing the stretch of the Transcontinental Telegraph from Fort Kearney to Salt Lake City were never able to sufficiently comprehend the Transcontinental Telegraph. This leads to the further assumption that Native Americans could not develop the technical acumen needed to surreptitiously tap, monitor, eavesdrop, or otherwise access the signal flow of the Transcontinental Telegraph and listen in to discern the content of telegrams. It is as if the wonderment of such a technological advance in long-distance communication was beyond the native ken, something spun from magic of a sort rather than the manifestation of a new communication technology with a purpose.[105] In one way, this is not so different from the attitude of telegraph corporations at the onset of the Civil War. The telegraph corporations insisted on creating and controlling the USMT rather than support a Signal Corps telegraph because the Union Army should leave telegraphy totally and completely in the hands of electricians rather than take on telegraphy as its own deployed military and security asset.

Arguments deriving from this attitude approach absurdity. The armaments manufacturers and suppliers for the military, such as Samuel Colt, never to my knowledge forcefully argued that military personnel should not own or use a Colt revolver and that instead such ownership and deployment of a Colt revolver in military action should remain in private hands, only owned and utilized by employees and officers of Colt's Manufacturing Company. The same could be said for all suppliers of ordnance, armaments, saddles, and so on. The idea that the Signal Corps would be incapable of understanding and deploying tactics, strategy, and doctrine about the telegraph, an idea promulgated during the Civil War by commercial telegraph operators to a great measure for their own commercial and financial success, is as nonsensical as arguing the army was incapable of learning and utilizing a new technology in armaments.

The fallacy of promulgating, believing, arguing, or implying there is a special, precious, unobtainable technological knowledge that presents an insurmountable obstacle to certain groups of people or individuals presumes limits on others' imagination. These limits are especially acute about the telegraph and the indigenous peoples of the Great Plains. Evidence is everywhere that Native Americans of the Great Plains, and throughout western North America, found no significant obstacles in developing sufficient technological mastery of a wide and various range of nonnative technologies from firearms to frying pans. Leaders and members of Great Plains tribes and other indigenous civilizations of western North America routinely visited the new towns and cities of white settlers, the emerging forts and stockades of the military, ranches and farms, trading posts, and, once they began to be built, telegraph stations. Indigenous peoples routinely observed the activities of white settlers and army troops and officers. Eugene Ware recalled visiting the Pawnee agency in fall 1863 and at first being taken aback during a meeting when "one Indian after another came, and stood at the window looking in" only to learn from others at the meeting that "the Indian thought by such attention he was doing us a great honor, and that we ought to consider we

The railroad, the telegraph, and the bison slaughter (Denver Public Library Digital Collections, call number Z-3803)

were being complimented."[106] Native Americans often were social with white settlers and were often observant of white settlers. Some indigenous leaders visited eastern North America—Chicago and other towns and cities—and sometimes visited with government officials in Washington. In those eastern metropoles, they met people, held meetings, and participated in events. They also stayed in hotels, ate in restaurants, traveled on railroads, and generally interacted with the new industrial technologies of nineteenth-century American urban life.

All this is well known. Yet a lingering, unvoiced assumption pervades that the nineteenth-century indigenous peoples of the North American frontier were incapable of understanding the telegraph beyond knowing it was a form of rapid long-distance communication invented by, used by, and understood only by the white man. This limited imagination—the inability to imagine that Native Americans may have understood far more about the telegraph and the necessary technical acumen needed to send and receive messages than the written record and archival residue indicate—is a limited imagination that also masks a potential manifestation of racism among those whose imaginations were and remain constrained in such a way. To imagine that not one indigenous person on all the Great Plains situated along the stretch of the Transcontinental Telegraph from Fort Kearney to Salt Lake City was capable of figuring out how to tap into the Transcontinental Telegraph and how to learn and understand Morse code, nary a one, begs credulity. This also is to imagine not one indigenous person closely observed the activities of a telegraph station. It is to imagine not one indigenous person was capable of recognizing that the process of sending and receiving a telegram was built on written messages and that Morse code was a form of written and aural language that could, with sufficient exposure and experience, be learned by observation. The dots and dashes of the telegraph signal constituted a telegram. A telegram was usually a message first written on a piece of paper, or a message first dictated orally to a telegraph operator.

There was also the daily routine of the signal flow on the Transcontinental Telegraph, which basically was to keep the signal flowing as much as possible. Signal flows are monitored in times of conflict, not only for the knowledge to be gained from the data content transmitted by the signal flow, but also for the knowledge to be gained from the patterns and routines of signal flow.[107] Anyone who closely monitored Transcontinental Telegraph signal flow every day, even without knowledge of Morse code, would have likely noticed on the afternoon of January 27, 1865, all signal flow came to a halt for a period of time. A message was then sent simultaneously to every conceivable station and location along a three-hundred-mile stretch of the line. This unusual signal flow was significant and was observable to anyone listening in, regardless of any knowledge of ciphers or codes. From the halt of all telegraph traffic, followed by a period of silence on the line, and then suddenly a signal flowing everywhere at once, an observer did not need to know the content of that message to realize the message was highly significant. The signal flow said it. The Great Prairie Fire confirmed it.

It is more likely to imagine that among the indigenous peoples of the Great Plains there were those who did, over time, eventually teach themselves, through observation, some rudiments of Morse code and how to tap a telegraph line. It is even more likely that some indigenous peoples of the Great Plains recognized, through observation and interaction, that the written messages and pieces of paper upon which telegrams were based were not always carefully secured and therefore could, on occasion, be read or taken away before or after a telegram had been sent. In August 1865 Pawnee scouts with army troops near the Powder River found an old Native American campsite strewn with "scraps of telegrams taken by them (warriors) at Julesburg" from earlier that same year.[108] In some cases, that may have been as easy as going through the trash pile of the telegraph station, particularly if that pile of paper was not secured and then burned daily.[109] It is easy to imagine that at telegraph stations, some telegrams given by oral dictation were overheard by visitors and observers, including indigenous peoples. The elusiveness of traditionally recognized and honored evidence in the way of archival documents, newspaper accounts, journals and diaries, or interviews confirming this possibility of indigenous knowledge of technology and the telegraph is not an excuse or license to limit the imagination in a way that risks promulgating a racist attitude toward indigenous knowledge of technology and the telegraph. The more responsible approach is to shed, disdain, and reject such imaginary limits and recognize the possibility—the probability—such indigenous knowledge of the telegraph did emerge and was effectively deployed.

In so doing, one can cast the extensive and ongoing attacks by Native Americans against the Transcontinental Telegraph in a more informed manner that recognizes the strong likelihood of a deep, rather than surface, knowledge held by the indigenous peoples of western North America of the power and potential of the telegraph. One also can take another step forward in debunking paternalistic and racist attitudes toward indigenous peoples and nonnative technologies, an attitude that for far too long has promulgated a racial hierarchy built and maintained upon beliefs, attitudes, and stereotypes of primitivism. An examination of the sophistication, speed, clarity, determination, and bravery of Plains peoples as they continually attacked the Transcontinental Telegraph in 1865 and 1866 cannot possibly yield a conclusion that Native Americans held some sort of technologically primitivistic notions about the telegraph. Instead, they clearly invented and put into practice the idea of an ongoing wave of denial of service attacks against the electronic communication network of an occupying power. Such attacks remain to this day a powerful, pervasive, and relevant military response to an enemy force.

3

Changes in the Forecast

Data Gathering, Mapping, and Weather Predictions

A storm in its gradual march along the coast will give notice itself of its advance along a chain of stations connected by telegraph, and charged with no other duty than that of watching it.

—*1874 Annual Report of the Chief Signal Officer*[1]

The year 1866 proved to be a watershed in the development of American telegraphy. The National Telegraph Act of 1866 allowed for stringing telegraph wire on post road routes, empowered the postmaster general to set telegraph rates for government use of commercial telegraph networks, and required that telegraph companies activating post road right-of-way uses could be bought out by Congress five years later.[2] The same year, Congress passed the Act to Increase and Fix the Military Peace Establishment of the United States.[3] This act permanently increased the size of the army, led to significant new troop deployments and fortifications in the territorial West, and reestablished the Signal Corps as the central unit in the army for military communication, including telegraphy. Albert Myer was restored to command of the Signal Corps.

During its years of operation, the United States Military Telegraph (USMT) Company had among its Washington telegraph operators George Maynard.[4] Arriving in Washington in 1864 from Michigan, where he had been a telegraph operator, Maynard was brought into the Washington USMT office. On April 14, 1865, Maynard was in the audience at Ford's Theater for the performance of *Our American Cousin* and became an eyewitness to the Lincoln assassination, his first of two close encounters with presidential assassinations. Maynard immediately made his way to the telegraph office at the

War Department and coordinated the nationwide sending of telegrams announcing the assassination and death of Lincoln. In 1881 Maynard loaned money to Charles Guiteau, who had been an occasional boarder at the home of Maynard's relatives in Ann Arbor, Michigan. Believing the money was, as Guiteau assured Maynard, to cover rent and other sundry expenses, Guiteau then used the borrowed money to purchase the pistol with which he assassinated President James Garfield.[5]

The Establishment of the Signal Service

Maynard was hired by Western Union for its Washington office after the war and was then recruited into the Signal Corps in 1870. In particular, Albert Myer recruited Maynard as part of the establishment of the Signal Service, the weather-observation branch of the Signal Corps authorized in 1870 by Congress. Maynard spent two years with the Signal Corps, traveling extensively while overseeing the establishment of a Signal Service telegraph network.[6] The Signal Service network was designed to collect weather data and route that data among Signal Service weather observation stations and to Signal Service meteorologists in Washington.[7] In particular, Maynard planned telegraph network routes and developed new traffic patterns and circuits for Signal Service telegraphy. This involved a range of tasks: overseeing new line construction, planning for the expansion of Signal Service telegraph lines and stations, and coordinating the integration of Signal Service and military telegraph lines with commercial telegraph lines and networks. The goal was rapid, regular, and reliable telegraphic data flow from Signal Service stations across the continent to and from other Signal Service stations, and to and from the main Signal Service meteorological office in Washington.[8]

The daily weather observations and forecasts of the Signal Service were a huge undertaking that enabled a massive expansion in military telegraph activities on the North American continent, especially in western North America. One report in 1884 compared the daily transmission of Signal Service weather observations to a "telegraphic roll-call of all the stations spread over the country."[9] This continental expansion of military telegraphy sparked an increase in military telegraph line construction, personnel, equipment, and transmitted military-related information that dwarfed the previous battlefield and temporary field telegraph construction activities of the USMT during and immediately after the Civil War. The USMT strung many miles of field telegraph lines during the conflict, but on occasion those field lines were temporary installations contingent on battlefields and troop movements.

Signal Service headquarters, Washington, DC (National Oceanic and Atmospheric Administration, National Weather Service Collection, image ID wea01313)

Other lines were commercial lines rebuilt by the USMT. Virtually all these field telegraph and rebuilt commercial lines were in eastern North America. Some new Signal Corps and Signal Service military telegraph lines and stations were built on the Atlantic Seaboard and around the Gulf Coast and Great Lakes, but the vast majority of new telegraph line and station construction for the Signal Service and the military telegraph took place west of the Mississippi River.

Codes, Ciphers, and Weather Observation Reports

In 1866, after being reinstated as commanding officer of the Signal Corps when the Civil War ended, Myer soon turned his attention to codes and ciphers. He argued that methods used during the recent conflict, such as variations and updates of the Alberti disk, would no longer be sustainably secure for future developments in military and government communication. Signal Corps officers found that the disks used in the war proved secure, but recent experiences and tests showed "when a copy of the disk is had, it cannot be relied upon to resist the efforts of persons who have made the subject a study, and made themselves masters of the different modes of deciphering."[10] Myer in particular noted that the rising "attention foreign powers now give to the subject of telegraphy" meant the Signal Corps would now need to "practice and experiment to test modes of cipher." The first significant advances in encryption deployed by Myer and the Signal Corps in the name of national security were not directed against foreign powers, however. They were instead deployed against the commercial telegraph companies, such as Western Union.

Congress passed a resolution in February 1870 authorizing the secretary of war "to provide for taking meteorological observations at the military stations and other points in the interior of the continent" and also issue storm warnings. In March 1870, in no small part because of lobbying by Myer, the task was assigned to the Signal Corps. Myer observed that the "duty thus imposed upon the Department was one for which the popular mind had been in some degree prepared by the recorded labors and theories of meteorologists." He added, "The undertaking upon a scale of such magnitude as that provided by the law, had not been generally contemplated even in this country."[11] There had been growing efforts to observe weather conditions and to use the telegraph for collecting data about weather patterns, including significant work by Joseph Henry and the Smithsonian Institution.[12] None of that previous work, however, was carried out to the scale, scope, and depth that Myer envisioned. Weather information was seen as a security issue and was prioritized by the military. Congress shared this view on weather and national security and also saw the Signal Service weather forecasting as supportive of commerce, particularly American agriculture. The move into weather forecasting led to a significant expansion of the Signal Corps into telegraphy, because the Signal Service used telegraph reports from across the states and territories to gather and relay daily information (often thrice daily) about local weather conditions and patterns.[13] These data were sent to Washington for analysis, and maps of weather patterns, weather forecasts and reports, and longer-range predictive forecasts were routinely issued. Myer saw the 1870 measure as an opportunity for the Signal Corps that would stimulate the expansion of budgets, troops, officers, and personnel.

As Myer noted in his 1870 annual report, the daily continentwide gathering, analysis, and dissemination of meteorological information was "not a subject for trivial contemplation" but a project "capable of indefinite extension" that, if successful, promised benefits "vast and lasting." The project "was a work to be entered upon with a sense of grave responsibility." The implementation of such a service included an array of decisions: what weather data to gather, who should do the data gathering, where the data should be gathered, how it should be tabulated and formulated, how often it should be gathered, where it was to be sent and how often, the need to consider the "mode of publication" and information dissemination, and, finally and most important, "the necessary arrangements for telegraphing the reports."[14] This included employing a suitable telegraph cipher for weather data to compress information and save costs. For accurate forecasts and particularly to warn of approaching storms, thrice-daily telegraph reports were recommended. This schedule claimed to allow no more than eight hours between updated weather reports. It was believed "it would be rare that a storm of magnitude would progress more than three hundred miles in that time." The times of observation took into account various periods throughout the day "when the telegraph wires are not so thronged as to prevent their use" in order to ensure the reliable flow of data. Systemwide coordination for gathering weather data meant "the observations shall be simultaneously had throughout the whole system of stations" with the "movement of the reports upon the telegraph wires" as "synchronous as can be made. The result will be to give thrice daily a synoptic view of the atmospheric conditions over a greater portion of the States and Territories of the United States." Myer knew this was ambitious and that nothing similar had previously been done "upon a scale of similar magnitude."[15] This was doubly true for both weather forecasting and for the signal flow along telegraph networks. These levels of data gathering and signal flow on telegraph networks, day in and day out and cast across as much of North America as possible, dwarfed even the largest wartime efforts of the USMT. The plan presented daily challenges of continuous signal flow timing, reliability, coordination, and simultaneity for telegrams and telegraph networks from a multitude of locations even beyond that of major users such as finance markets or the Associated Press.

Myer acknowledged that the "telegraphic transmission of the regular reports has presented a problem difficult of solution" and therefore the work of Maynard in routing signal flows and circuits was crucial. A circuit plan involving Western Union, International Ocean Cable Company, Northwestern Telegraph Company, and other commercial telegraph firms yielded seventeen separate routing and circuit configurations to ensure "the telegraphic labor needed for the movement of the messages of each group; for the exchange of message reports between different groups; between different places in different groups; and finally, for the assembling of all the dispatches in Washington." In some ways, this plan faintly resembled the emergence of the internet a century later in its 1960s–1970s era of ARPANET and MILNET "command and control" for secure communication and signal flow, particularly with versions of redundancy and distributed (rather than centralized) signal flow and routing. The ability of any individual Signal Service station to hold, redirect, or retransmit a message along the network was, basically, not altogether different from packet switching on ARPANET and MILNET. Indeed, the need Myer asserted to repeat, redirect, and retransmit many encrypted weather messages to ensure accuracy would eventually be a point of serious contention between the Signal Service and Western Union.

Myer if anything understated the plan: "I am not aware that a style similar to this has been before adopted."[16] For good measure, he opined that the ability of this network to expand and include more telegraphic reports was ultimately global. Foreseeing a continuation of growth in the number of undersea transcontinental cables, Myer found "the idea of a world-wide system of telegraphic weather reports is not as chimerical to-day as was thirty years ago the workings of the electric telegraph itself."[17] The telegraph companies agreed to provide four months of free service at the inception of this plan but declined to set rates for their services until having gained some day-to-day experience with the Signal Service nationwide signal flow.[18] The network commenced operations on November 1, 1870.[19] Myer noted in his annual report that on "November 1, 1870, at 7:35 a.m., the first systematized synchronous meteoric reports ever taken in the United States were read from the observer-sergeants of the signal service at twenty-four stations, and placed upon the telegraphic wires for transmission."[20]

The Signal Service telegraph network—and the signal flow—expanded in 1871. Greater focus on issuing cautionary signals warning of impending storms in an area now meant additional hours, including telegraph network hours, "by the fact that the station now must be ready to receive the telegraphic orders at night as well as in the day." Telegraph companies became less cooperative, which Myer attributed to a "misunderstanding as to the powers of the United States in relation to the telegraphic lines" of the commercial operators and their legal obligations under the Telegraph Act of 1866. Western Union was the culprit, suspending its network arrangements with the Signal Service for a time early in 1871.[21] But "the Franklin, the Atlantic and Pacific, and the Pacific and Atlantic Telegraphic Companies tendered their services promptly" as did the International Ocean Cable Telegraph Company with a "spirit of accommodation."[22] Although relations with the telegraph companies were "satisfactory," nevertheless "embarrassments" occurred from time to time, which Myer believed "cannot be remedied until the United States secures, by some arrangement, the absolute control of the lines extending over its territory."[23] Whether this meant nationalization, a new wholly-owned-and-operated government telegraph network, or something else to ensure "absolute control" Myer did not say. He did not provide a potential solution but pointed to a perceived need. Myer gushed about the continental-to-global possibilities of network expansion, calling for a Pacific cable, reporting a weather observer had been sent by the Signal Service on the 1871 *Polaris* north polar expedition, and musing on the possibilities of weather data from "Pembina, Sitka, [and] the Sandwich Islands."[24]

In 1872 Myer introduced a new cipher for the Signal Service. He and his cryptographers had constructed a new cipher that was shorter in words to reduce costs because commercial telegraph rates were based on the number of words in a telegram. The new cipher caused consternation for Western Union. Contrary to the assurances of Myer and the Signal Corps, the new Signal Service cipher introduced by Myer, despite requiring fewer code words than the previous Signal Service cipher, did not reduce transmission time but actually increased transmission time. The new cipher raised the expenses of Western Union (and other commercial firms) in carrying Signal Service messages in that it increased labor, because Western Union found the cipher used words "nearly so alike telegraphically that to use them to represent different groups of figures is to invite error, and to render accuracy absolutely impossible." For Western Union, the 1872 cipher was "so devised" to increase the amount of time and labor needed to accurately transmit Signal Service telegrams and therefore "reduce on the basis of a computation per word the compensation for the service."[25] Myer and his cryptographers had structured a new cipher that was shorter in words to reduce costs. The use of the new cipher with its shorter total of words per message simultaneously increased Western Union errors and thus demanded additional Western Union labor to correct those errors. This increased the amount of time the Signal Service messages occupied the Western Union telegraph lines on a daily basis. In essence, Myer found a way to pay less for more time on the American commercial telegraph networks, a neat trick that would be the envy and goal of many a data programmer to come in the future.

The 1872 cipher had fewer characters and code words than the 1871 cipher, thus reducing costs, because billing was on the basis of individual words (or on the number of coded word-units in a message). However, the 1872 cipher, while shorter on paper, was so to speak longer on the ear. Put another way, when the cipher was written for transmission and relay on the telegraph networks, the composer of the written message could see that the new cipher saved characters and words and thus saved money. When the telegraphers took the written message and relayed the message, the conversion from a static written text to a flowing aural-tactile text with telegraphers listening and simultaneously keying the message through the Signal Service network increased the likelihood of error during relay and transmission on the commercial telegraph networks.[26] Because the transition from a static written text to a flowing aural-tactile text increased the possibility of error, the new cipher took longer to successfully traverse those commercial networks once the message was no longer in written form.

The 1872 cipher was a form of data compression, because in comparison with the 1871 cipher, the 1872 cipher conveyed more data and information with fewer words and thus with lower costs for the federal government. Paradoxically, the successful data compression did not result in bandwidth compression. There was no reduction of bandwidth in deploying the 1872 cipher, nor was there a comparative reduction of time needed to successfully transmit a given amount of data. The continual need for error resolution by Western Union—after all, the commercial telegraph companies only received compensation from clients for the successful delivery of a correctly worded telegram—increased the number of required hours for transmission. Alas, those increased hours dedicated to error resolution were bereft of compensation for Western Union and the other commercial telegraph companies. The stunning result of the 1872 cipher was data compression that needed more time on the bandwidth. Data compression by the Signal Service forced a longer temporal duration of data dissemination and signal flow for Western Union. This was not because the 1872 cipher was kept a secret from Western Union. Western Union knew the cipher and had the required code books. The cipher was secure, not secret. The challenge was not in keeping secrets, but in keeping security by accurately discerning without error between very different words that sounded very much alike to the ears of Western Union telegraphers.

Western Union president William Orton was among those stunned by the "less data needs more time on the network" paradox of the 1872 cipher, attributing the paradox to choices and decisions about code words within the cipher. Orton testified that on the one hand, a more efficient cipher was of course acceptable, noting that the 1872 cipher was built "to convey more matter in a word. We have no objection to that." The problem was not in any single individual code word, but in any Signal Service telegram as a ciphered message consisting of a string of code words "because of the greater difficulty in sending words that are so liable to be misunderstood."[27] Very similar individual code words often had distinctly different meanings. Furthermore, the total number of code words was voluminous: a distinct word for every different tenth of a degree of temperature, for every millibar of barometric pressure, for wind directions and speeds, for cloud formations, for morning, afternoon, and evening observations, on and on.[28]

The individual code words often were indistinguishable even to the ears of seasoned telegraphers. Orton informed the House Appropriations Committee that the 1872 cipher contained many words that risked "a great liability to error . . . Tale, Tall, Tell; Them, Then; Real, Ream; Rid, Ride; Star, Stir; Swam, Swim; Care, Case; Hymn, Hymen; Cam, Came; Fame, Fan, Far; Finder, Finger; Fit, Fat; Flag, Flog, Flash, Flask; Flora, Floral; Fore, Fork; Tank, Tang, Ting, Tong; Fund, Funk; Tom, Tome; Tip, Top." Orton also reported the 1872 cipher occasionally called for a space between Morse code characters within a ciphered word, and a space between Morse code characters within a word was notoriously difficult for a telegrapher to hear and then accurately rekey for relay.[29] There also was the question of context in a Signal Service weather report telegram. Any string of code words in the cipher meant the telegram had no context to which the telegrapher could mentally refer. The Signal Service weather data telegrams sent from individual stations on the Signal Service network were nothing but a string of code words: no sentences, no phrases, no grammatical structure, no narrative, and therefore no linguistic context. Orton explained that "there is no telegrapher who does not understand the almost impossibility of distinguishing between such words as 'tip' and 'top' and 'flip' and 'flop.' All the difference in the signals is the space between the two dots." This meant "you have nothing in the context to illustrate which it means, or to contribute to settle the doubt. If the receiving operator understands that it is 'tip,' he puts it down so. It makes just as good sense as any other word—the words themselves conveying no meaning to the ear or mind of the operator."[30] The 1872 cipher conveyed meaning without context.

The differences of words and ciphered word-units which the eyes of the Signal Service and Signal Corps personnel could read with accuracy and ease became differences extraordinarily difficult to accurately and easily discern for Western Union telegraph operators simultaneously using their ears to listen and their hands to key and relay. The cipher exploited the perceptual differences between a written message and that same message heard and simultaneously keyed for relay through a telegraph network on the route from sender to recipient. Standard telegraph network practices of the day did not (yet) transmit a telegram encrypted once and only once into written form, which then traversed the system in written form ("print" telegraphy relayed entirely in print from sender to receiver). Practices of the day depended on the ears and the hands of the telegraphers, who heard and simultaneously keyed the messages for network relay. This difference between the cipher as written and the cipher as heard contradicts the assumption that a shorter code is often the most efficient code throughout a communication system. This also indicates that determining efficiency in a communication system can be a contested discourse between parties.

Another result of the 1872 cipher with its tendency for error on the part of Western Union was that the need to repeat messages also created the conditions for Signal Service operators to routinely "chat" with each other over Western Union lines as they checked and rechecked ciphered weather data for accuracy. This habit was not confined to Signal Service telegraph officers on commercial services. In 1883 the secretary of war reminded army officers and personnel that "The Government has no more become a 'public carrier' by building military telegraph lines than it has become a common carrier by owning a large number of quartermaster's wagons."[31] One might see Myer and the Signal Service as crypto-hackers, exploiting the weaknesses and inconsistences of a commercial electronic communication network. For Myer and the Signal Service, their exploitations were completely supported by Congress, the military, and the federal government.[32]

In sum, the Signal Corps created a cipher that was simultaneously shorter for the eye and longer for the ear and hand. The reduced telegram rate for the federal government was based on the length of the message in written form, and not on the time and labor (including error correction) of the transmission.[33] Western Union noted that systemwide daily tests of the new cipher on Western Union wires showed a reduction in words of approximately 31 percent, but an increase in transmission and relay time of 3 hours and 16 minutes. The old cipher used (on average) 7,876 words per day, occupying 10 hours 38 minutes transmission time, meaning the average rate of transmission per minute was 12.3 words. The new cipher used (on average) 5,370 words per day and took 13 hours and 54 minutes for an average rate of transmission per minute of 6.4 words. The 1872 cipher, used daily, averaged 3 hours 16 minutes more of Western Union network time with 2,506 fewer words than the 1871 cipher.[34] The Signal Service, like all government agencies, was billed for each telegram on a per-word rate (cipher or plaintext).

Drops, Circuit Routes, Metadata, Net Neutrality, and Hypothetical War

The Signal Service successfully argued that additional copies of the same Signal Service telegram, known at that time as a "drop," should not produce additional billing charges. Drops were routine for the Signal Service, and in some cases, an individual Signal Service telegram to Washington also would be flagged to add twenty-five to thirty drops across the Signal Service network, all at no additional charge. The Signal Service also successfully claimed the right to determine Western Union nationwide circuit reconfiguration. This allowed the Signal Service to determine the routing and flow of messages across the combined Western Union–Signal Service nationwide network interconnecting military telegraph lines with commercial telegraph lines, thus shaping nationwide traffic patterns with Signal Service control over Western Union and other commercial network switching. Control was justified on the basis of ensuring the timely delivery of Signal Service telegrams throughout any array of Signal Service observer stations (usually as drops) and on to Washington. Signal Service and Signal Corps personnel routinely did circuit reconfiguration of the commercial telegraph networks in the 1870s.

Another important issue regarding signal flow was, it might be said, a nineteenth-century version of the contemporary debate about net neutrality. Should the signal flow of one party take precedence over the signal flow of another party? Should commercial telegraph corporations be either allowed the flexibility or compelled by the government to favor one customer over another in establishing signal flow, priority of network access, and commercial rates? For government uses of commercial telegraph networks, could commercial corporations be required to provide the federal government with rates and a fee structure at a lower cost than other customers (yes, according to the provisions of the Telegraph Act of 1866) and could the federal government, particularly the military, have both the power of priority access and the power to reconfigure circuits and signal flows on commercial telegraph networks as it saw fit to meet national security needs? Congressional hearings in the 1870s indicated that the answers to these questions on priority access for the military over commercial companies and on circuit reconfiguration and signal flow control in the name of national security were "yes."

"Yes" was also the congressional response on the rights of the military to require its codes and ciphers for message transmission over commercial services. The right held for cases in which the codes and ciphers introduced errors into the relay of messages that required the companies to increase the number of hours each day that the military reconfigured their commercial circuits and signal flow without additional financial compensation in order to resolve the message errors the military-mandated codes and ciphers themselves had introduced. Congress agreed the Signal Service had the right, when necessary to ensure timely signal flow and data collection, to reconfigure commercially owned and operated telegraph circuits and networks.

As a byproduct of ceding control over routing, data flow, and circuit reconfiguration, the commercial services were somewhat compelled to share with the Signal Corps the significant needs (number of telegrams, distribution patterns, circuit flow, and similar) of major clients, such as the Associated Press. This meant Western Union and other commercial services needed to inform the Signal Service and the government about the daily usage habits and needs of its private clients without necessarily gaining the prior consent or knowledge of those clients. This was, in effect, a nineteenth-century case of commercial telecommunication corporations quietly divulging to the federal government private client telecommunication metadata without the prior knowledge, awareness, or consent of that private client. In the end, this was accomplished through ongoing negotiations with the army, the commercial services, and on occasion Congress. In so doing, it appears that Western Union and other commercial telegraph services avoided significant government regulation that would have ensured these government abilities by force of law. This also suggests that the quiet relationship of cooperation in the name of national security between the federal government and private telecommunication and information firms reaches back to the telegraph, including the deployment of the telegraph in western North America for weather data collected and disseminated by the Signal Service.

In the name of efficiency and cost reductions, Myer fragmented the embodiment of the telegrapher and demonstrated that electronic media technologies were capable of fragmenting the human senses of media perception and reception, exploiting differences in human sensory perception and human motor skills to the benefit of the programmer. Thus the Signal Service introduced a significantly advanced system of telegraphic message encryption at the same time that it instituted a tremendous increase in the number of telegrams across its network of observer stations and also a tremendous increase in the number of telegrams to Washington. In short, the Signal Service delivered higher levels of data security and of data flow for North American weather in the name of national security and agricultural productivity, enabled by careful coercion over the commercial telegraph companies. The need to wire into nature and constantly discern the climate of North America—the daily gathering and synoptic analysis of weather data—shaped the conditions whereby the government advanced a discreet relationship with the commercial telegraph companies. This enabled government influence over the daily practices of those firms, justified in the name of national security. As a result, Myer and the Signal Corps found their expansion of Signal Service telegrams and increasing lines and interconnectivity positioned Myer and his unit more and more "at the center of an electric intelligence network, a giant web spanning the nation."[35]

Myer, the Signal Service, and the Signal Corps were supported by both Congress and the secretary of war in their 1872 confrontation with Western Union over ciphers, circuits, and costs. In June the secretary of war issued a directive to the Signal Corps and Signal Service, but written directly for the commercial telegraph companies, that specified a wide range of instances requiring government rates for commercial firms.[36] By 1875, Myer confidently reported that the success of the Signal Service weather network meant that "the information from the office of the Chief Signal Officer reaches daily one-third (1/3) of all the households in the United States."[37]

Another example of national security, signal flow, and the state secret lurked in various discussions between the commercial services, the military, and Congress on the hypothetical possibility of future hostilities and the subsequent conditions that war might warrant for the telegraph. Although most parties presumably saw the conversation as hypothetical, a few parties likely saw the conversation as pragmatic: perhaps even a present condition rather than a future possibility. One such individual was arguably General Albert Myer.[38] Myer reconstituted the Signal Corps in 1866 in the aftermath of the Civil War. Myer saw to it the Signal Corps was the main unit of all communication practices of the army and also saw to it the Signal Corps was present and active in the Trans-Mississippi West and present and active at an increasing number of western forts and stockades during the long period of conflict between the army and Native Americans. Myer had first field-tested his military theories and applications of signal communication during the conflicts between the army and the Navajos prior to the Civil War. Myer knew that telegrams and messages on conflicts between Native Americans and the army, conflicts between Native Americans and state or territorial militia, and conflicts between Native Americans and white settlers routinely flowed as a part of Signal Corps telegraph traffic and communication throughout western forts and stockades and on to Washington. Myer knew that the general command of the army (William Tecumseh Sherman and additional officers in high command) were familiar with this message traffic. He knew the army increasingly used the telegraph in western North America as a tactical asset in Native American surveillance, containment, and conflict. Myer also knew that his latest creation, the Signal Service and weather information, was considered both a public service and a national security measure, and that weather information could be of value in military matters, such as coordinating army campaigns against Native Americans.

Electronic Surveillance and National Security

Put in twenty-first-century terms: did Myer conceptualize the army, and by extension, the federal government, in a state of continuous warfare? Did he perceive this as long-term asymmetric counterinsurgency against nonstate actors? Did he believe that part of waging this continuous unconventional war against nonstate actors was the building of Total Information Awareness? Did he believe this meant that commercial telegraph systems were required to prioritize Signal Corps and Signal Service telegraph activities, and commercial telegraph systems were required to allow the Signal Corps and Signal Service to dictate routes and circuits? Were commercial telegraph systems required to share any and all knowledge and data about system operations, patterns of usage, billing and revenue information, construction and operation costs, and the usage habits of other customers (such as the Associated Press) with the Signal Service, the Signal Corps, the army, and the federal government? In other words: Did Myer see the relationship between the military and the commercial telegraph systems in a nineteenth-century manner akin to those in the twenty-first century who argue that the Global War on Terrorism justifies such an expansive range of government activities, monitoring, and surveillance on electronic information networks? Although these questions about Myer cannot be answered in the literal sense, they are offered herein to suggest—cautiously—that certain views on the power of the federal government over electronic information networks during times of conflict, counterinsurgency, or an ongoing war on terrorism may have historical roots in the restoration of the Signal Corps after the Civil War. This included the Signal Corps doctrine and approach to the deployment of signal communication and the telegraph in ongoing and continuous conflicts between the army and Native Americans, border control with Mexico, surveillance activities, and the Signal Service daily procedures of gathering, encoding, and transmitting weather data by telegraph. In terms of media archeology, the Signal Corps is a reliquary of contemporary cyberwarfare.[39]

Looking forward into the twenty-first century provides one possible perspective on Myer and the Signal Corps. Looking back at early modern European history may provide another perspective: Myer as a version of Jean-Baptiste Colbert, the information master of Louis XIV.[40] Colbert sought to acquire and control information in order that he could most effectively serve Louis XIV, and in the process Colbert built libraries and archives as well as a surveillance system. Although Myer was not comparable in the sense of one person mastering information in allegiance to a sole sovereign, Myer and the Signal Corps were a part of the ever-increasing data gathering and information mastery of the government. The Signal Service, with weather information gathered across the continent and distributed to Washington for analysis and dissemination, represented a system for gathering and mastering information.[41] Thinking back to Edwin Stanton, it might be said that Myer adapted from Stanton the information mastery of the USMT as well as the mastery of data collection, analysis, and control. That the Signal Service on occasion conducted surveillance of urban labor unrest for presidential administrations suggested the potential intimacy of government data gathering and government surveillance.

On occasion, Myer used his telegraphic surveillance skills in an effort to enhance his personal relations with fellow army generals. When Nellie Grant and her new husband sailed on their honeymoon to England, upon their arrival Myer "sent Grant a cablegram at 3 A.M.," assuring Ulysses Grant of safe passage for his daughter. Myer also sent Grant daily nationwide weather reports even when Grant was at the seashore for the summer. Grant found this personal touch of surveillance and data irritating.[42] This small tale, while at first glance perhaps amusing, indicated the potential for personal power and heightened influence suggested by increasing skills and abilities in government surveillance. In behaving like a sycophant, Myer also demonstrated the ability of his office to bring together electronic information networks, transportation schedules, weather reports, and on-site observations to profile the activities of one person. Personal profiling enhanced by telegraphic surveillance already was underway for certain Native Americans who refused to stay on a reservation and for known or suspected criminals. With Nellie Grant, Myer demonstrated that the same monitoring, surveillance, and data-gathering techniques could be applied to anyone who might be considered, for any reason, a person of interest. The uses of the telegraph for surveillance, monitoring, pursuit, and extradition of suspected criminals and persons of interest expanded rapidly in the 1880s, becoming a standard practice of the Pinkerton National Detective Agency.[43]

The 1877 Railroad Strikes

The Signal Service, Signal Corps, army, and President Rutherford Hayes all worked together during the 1877 Railroad Strikes to implement a nationwide system of surveillance and monitoring that centrally included the telegraph. Signal Service personnel in striking cities and cities with indications of labor unrest routinely infiltrated labor meetings and strike-related events. They did not appear in uniform but rather in civilian clothes and reported their observations and findings via telegraph back to Washington. The telegrams were usually sent directly to Myer.[44] The telegrams and dispatches were relayed to Benjamin Montgomery at the White House Telegraph Room and reached Hayes, his cabinet, and military commanders every day.[45] Some reports revealed the ongoing practice of synoptic surveillance. For example, on July 26, 1877, Secretary of War George McCrary briefed Hayes and the cabinet on recent strike developments using information gleaned from various Signal Service telegraph reports. The *New York Times* noted that the reports were possible because of "instructions having been given Signal Service observers at distributed points to make frequent and detailed reports. These reports show that the strike is rapidly extending, and there is now scarcely a road in the Middle and Western States that is not blockaded."[46] Hayes and McCrary visited the night practice of the Washington-area Signal Service in early August. They witnessed signal communication by torch signals, mortar shells, and telegraph. The night practice closed with a "Presidential salute of 21 shells" and Hayes complimented the Signal Service on its actions "during the recent troubles."[47] In a retrospective on the Hayes administration published in March 1881, the *New York Times* recalled that during the 1877 strikes Hayes "was in immediate telegraphic communication with every important point in the country" and used "the Signal Service as a corps of riot observers. Its members were directed to report facts, and facts only." Explaining the common ground of synoptic meteorology and synoptic political surveillance, the *Times* told readers that the labor strikes were under continual observation and analysis via telegraph at local, state, and national levels. "The progress of disturbance or of its premonition was mapped and tabulated as the intimations of a storm would have been, and the plan was followed of moving troops so as to stop the advance of rioting rather than to follow in its wake."[48]

The White House was no longer an island isolated from telegraphic connectivity, as had been the case for Abraham Lincoln. Rutherford Hayes, the Signal Corps and Signal Service, and the army had now taken the first steps needed for the White House to eventually become a secure command at the nexus of American electronic communication networks—the beginnings of a White House Situation Room. The sociotechnical system developed to continuously monitor North American weather and nature with a synoptic vision was adapted to continuously monitor dissent and political unrest with a synoptic vision. American government surveillance tactics and strategies involving the telegraph in the 1870s and continuing to involve new electronic

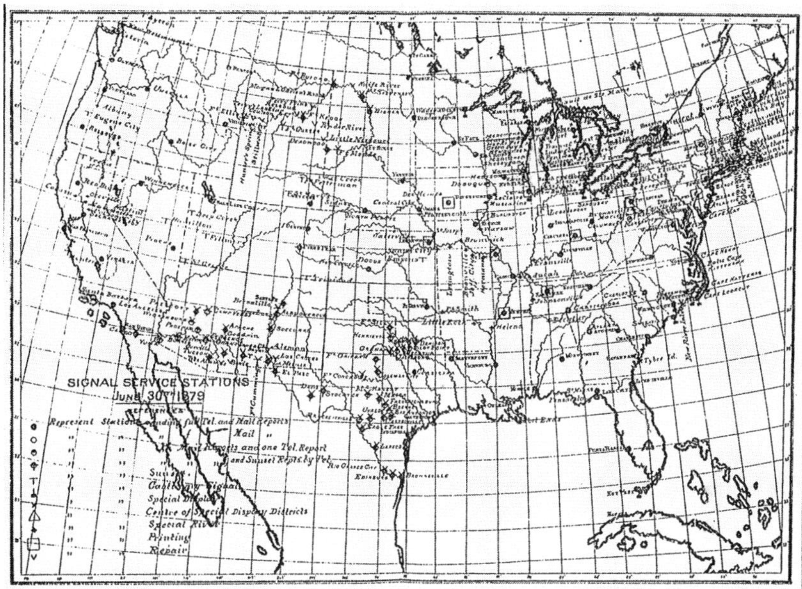

An 1879 map of Signal Service weather observation stations (*1879 Annual Report of the Chief Signal Officer*, p. 166)

communication networks up to the present were shaped by a growing mastery of knowledge about North American nature.

The surveillance by the Signal Service and subsequent actions of the army and the White House in response to the 1877 Railroad Strikes were significant departures from previous practices. Ulysses Grant, now out of the White House, spoke out with disdain. Grant remembered that the Republican press "thought it horrible to keep U.S. troops stationed in the southern states" to protect the lives of African Americans and the "sound of indignation belched forth" for years. After so much hand-wringing about protecting lives with federal troops, now with the 1877 strikes there was loud support for using troops to protect property, with "no hesitation about exhausting the whole power of the government to suppress a strike on the slightest intimation that danger threatens."[49] Myer was not Colbert in the sense that Myer did not uniquely embody (with the blessing of the sovereign) the information mastery and surveillance abilities of a nation-state. Nor was Myer like Stanton, exercising power in the extraordinary circumstances of the Civil War. Yet Myer and the Signal Corps did represent and exemplify the growing tendencies and skills of government individuals and agencies—military and civilian—toward information mastery. To a degree, Myer normalized the Civil War information mastery of Stanton and the USMT. Extraordinary circumstances for the telegraph in the recent American moment of existential crisis were selectively transformed into short-term tactical actions and long-term strategic doctrines of American national security after the existential crisis, transformed into standard surveillance practices during times of relative peace.

By comparison with the major nation-states of Europe, information mastery and government surveillance came late to the United States. These government information-gathering and state-surveillance activities gestated in Europe in the mid-seventeenth century and did not gestate in the United States to any great degree until the mid-nineteenth century. Given the moment of American gestation, it was to be expected that the information mastery and surveillance possibilities of the telegraph would arise in the visions of some American military officers such as Myer—and probably also in the visions of some American commercial telegraph executives such as Western Union president William Orton. In this sense, it was possible that Myer and Orton saw eye to eye. That is not unusual among adversaries. The state secret issues in nineteenth-century electronic communication networks and national security are visible: the tactics and strategies of military communication, the rise of a surveillance society in the United States, and the theories, doctrines, and hypotheses of what war, conflict, and political dissent would mean to the nation's systems of communication, including the telegraph. Experiences with surveillance of weather and nature were among the influential factors shaping civil and military surveillance.

Congress supported the Signal Corps and Signal Service on the issue of encryption and told Western Union that additional services to ensure the accuracy of transmitting Signal Corps ciphers, including the routine repeating of cipher messages to resolve transmission and relay errors, were not subject to additional billing charges. Congress and the War Department somewhat increased the Signal Corps budget during the latter decades of the nineteenth century, although military telegraph line construction was usually carried out by regular troops under Signal Corps supervision. Nevertheless, as a general condition, the Signal Corps remained somewhat underfunded. Whatever budget increases ensued paid for more Signal Corps soldiers and officers, advanced training, new experimental technologies, weather research, and thousands of miles of Signal Corps telegraph lines, stations, and equipment along seacoasts and the Great Lakes across western North America and Alaska. The funds also supported Signal Service and Signal Corps Arctic Polar Expeditions in the 1880s. (A further discussion of Alaska and the Arctic is in chapter 4 of this book.)

Finally, in order to fully carry out the mandate of the Signal Service and also to increase the probabilities of accurate weather forecasting, the Signal Service telegraph data needed to flow not merely across the states and not merely across the states and trans-Mississippi territories. It needed to flow not only nationally, it needed to flow continentally (and, by the dawn of the twentieth century, it then also needed to span a new empire with territory across the planet). In order for the signal flow of American national security to become the ubiquitous global phenomenon it is today, the signal flow first—and beyond all else—needed to conquer the North American frontier. That meant gathering many forms of knowledge from across the entire continent for dissemination to Washington and subsequent data storage, retrieval, analysis, and application. Thus, the theme of signal flow encompasses nineteenth-century manifestations of contemporary issues such as global power, net neutrality, metadata, and encryption. The emergent secure command enhanced the ability of the military-government-intelligence system to dictate and enforce—discreetly, of course—the national security needs of the high ground, signal flow, and state secrets for commercial telegraph networks and services.

Old Media, New Media: The Telegraph and Telephone in Washington

Despite the implications of new encrypted data flow and greater data flow for Washington suggested by the work of the Signal Service, electronic communication security in nineteenth-century Washington remained inconsistent. A number of problems were evident, for example, at the Capitol: the security of telegrams, locations of telegraph offices, and vulnerabilities of overhead or aerial (rather than underground) telegraph and telephone wires were becoming clear. On May 9, 1874, Architect of the Capitol Edward Clark responded to House Speaker James G. Blaine on questions about the commercial telegraph offices and stations in the corridors of the South Wing of the Capitol building.[50] Telegraph companies had established corridor offices within the Capitol to serve a range of clientele, including Congress, the Supreme Court, Library of Congress staff, other government personnel, reporters, and visitors. Blaine had asked Clark to recommend ways in which the "telegraph instruments" could be "so isolated that it shall be impossible for any unauthorized person to hear and obtain messages."

Clark consulted with Western Union and Franklin Telegraph, both of whom ran commercial telegraph offices in the corridors. He also sent J. F. Knapp, the operator of the government telegraph for the Capitol South Wing, into the field to find out where the telegrams sent by the stations in the Capitol corridors went as they left the Capitol for relay on various lines and networks. Knapp traced message flow from the Capitol up the Eastern Seaboard to Boston. He found that many messages went from the Capitol directly to the telegraph operators at Willard's Hotel in Washington. Although the Willard's telegraph men were considered reliable, where a telegram was relayed after that and whose hands and ears it passed through was indeterminable.[51] Knapp, giving advice echoed by individuals from the commercial telegraph services, told Clark that isolation was effectively accomplished by equipping the corridor stations with two items. Knapp recommended an ear trumpet that captured and directed the sound of the key only to the operator's ear and screens around the telegraph key that prevented the possibility of seeing the hand of the operator as he worked the key. Knapp called these security upgrades "silent instruments." Clark also was advised that a more secure system might allow the telegraph operator to remain in public view, but with the hand on the key masked by a screen so as to be unseen by observers. A telegraph operator in a secluded room "might allow people inside his office unobserved; whereas situated in a corridor, the office is so public that an operator would not dare to do such a thing, as it would be too readily observed." In short, Clark was advised to mute the key, mask the hand, and keep the telegraph operator in public view. This maintained the privacy of the telegram while keeping the telegraph operator in a panopticon. Contemporary social media texters routinely experience this tension of exchanging private messages in public view.

Edward Clark, the first individual to hold the title of architect of the Capitol, took significant steps to ensure telegraphy (and telephony) were secure electronic communication systems for the Capitol and for other federal buildings in the District throughout his tenure, an imperative acted on repeatedly by subsequent architects of the Capitol. From 1865 to 1902, Clark oversaw many projects, including a complete reconfiguration of telegraph and telephone connectivity after the 1898 gas explosion in the Capitol. He reconfigured internal telephone and telegraph wires, improved privacy, and installed state-of-the-art switching and receiving equipment.[52] In 1888, the commissioners of the District of Columbia, the superintendent of buildings and grounds, and the army chief engineer helped bring together a plan for moving the telegraph and telephone wires (and many circuits and switches) of Washington and nearby areas underground. Even at that date, there was not yet an expansive telegraph and telephone network or wire and switching infrastructure for federal buildings in the district owned and operated by the government. The only government connectivity not run through commercial networks (and overhead wires) were two underground telegraph cables running from the State, War, and Navy Building through the White House to the Capitol. Commissioners also planned to put wires for alarm systems and local police telephone boxes underground.[53]

Edward Clark and others in Washington also dealt with the arrival of a new electronic communication technology: the telephone. In 1877 George Maynard built and operated the first telephone exchange in the federal district.[54] Myer and the Signal Corps were among the first clients. The White House also was an early client, where Maynard installed a telephone as well as a number of electrical devices, including a doorbell. Other government offices, local businesses, President of the District of Columbia Board of Commissioners Josiah Dent, Gardiner Greene Hubbard of Bell Telephone, grocer John Magruder, and Spencer Baird of the Smithsonian also subscribed. Maynard also represented the Edison Phonograph Company. Although many government agencies and businesses were listed in the directory exchange published by Maynard in 1879, the telephones of Myer and the Signal Corps offices were not published.[55] This suggests that public knowledge of Washington telephone subscribers and network connectivity, in this and likely other cases, was incomplete.

On the other hand, the nascent Washington telephone exchange brought the telephone to central locales and institutions of Washington society and public life, in some instances by networking what might be called the invisible nodes of the network with the visible nodes of the network. For example, Maynard ran a telephone line from a Signal Corps office at 17th and C Streets in a connection that bypassed the White House, bordered Lafayette Square, and terminated at Wormley's Hotel at 15th and H Streets, with an extension line from Wormley's to Welcker's Hotel and Restaurant just down 15th Street.[56] James Wormley, the proprietor of Wormley's Hotel, was an extraordinarily successful African-American hotelier and entrepreneur in Washington. In addition to the hotel, he ran successful operations in horse-drawn carriages, stables, catering, and rooming houses and supported educational and civic initiatives as well. Wormley's was a favorite of European delegations, and the hotel hosted many dinner meetings, including meetings about the disputed 1876 presidential election. Wormley apparently attended the 1876 Centennial Exposition in Philadelphia and is believed to have seen the telephone demonstrations of Alexander Graham Bell. He was the thirteenth subscriber to George Maynard's Washington telephone exchange.[57] Wormley and his family also owned and operated a number of boardinghouses and businesses on or near the 1500 block of I Street NW, including a grocery store at 1527 I Street NW. One Wormley boarder before and during the first months of the Civil War was General Winfield Scott.[58] A number of Signal Corps officers stationed in Washington in the 1870s and 1880s lived in the vicinity, possibly at a Wormley property on occasion. It is likely that these officers frequently patronized Wormley properties and had meals provided by the catering services. Albert Myer lived on I Street and at one point may have boarded in a Wormley property, and William Babcock Hazen, his successor as Signal Corps commander, later owned a nearby mansion at 16th and K Streets NW.[59]

George Maynard experienced the growing complexity of local market competition as well as the long-term consolidation of the Bell system during his years operating a Washington telephone exchange. The visions of Theodore Vail were among the factors leading to Maynard's sale of his Washington telephone exchange in 1880. The Maynard exchange eventually became part of the Chesapeake and Potomac Telephone Company, the local Washington-area Bell System service before the 1982 American Telephone & Telegraph (AT&T) divestiture. Vail was known by the Signal Corps before his career with AT&T. In his 1877 annual report, Myer praised the support and efficiency of the Railway Mail Service, where Vail was then general superintendent, for its "careful and energetic aid" in transporting Signal Corps and Signal Service instruments and equipment to "every portion of the United States."[60]

Maynard stayed on and thrived in Washington, continuing to run his electrical store and provide services. In 1894 Maynard was the key figure in organizing the Telegraphic Historical Society of North America and cohosted the first meeting of the society, held in Washington.[61] The new society, of which Maynard was secretary and treasurer, had a growing interest in the history, preservation, and collection of telegraph artifacts. In the early twentieth century, Maynard institutionalized these interests and activities at the Smithsonian, where he served until retirement in 1918, eventually becoming the curator of Mechanical Technology.[62] Maynard did not live long enough to see his spouse, Anne Belle Rogers Minor, ascend in 1920 to the presidency of the National Society of the Daughters of the American Revolution.[63] Yet by the time of his death, Maynard had become a significant individual in Washington: a prime figure in the development and deployment of telegraphy and telephony for the city, the federal government, and the District of

Columbia. George Maynard helped shape electronic communication security for the city and by extension the entire nation.

Clouds Give Way to Clear Skies

As communication security for the telegraph and telephone in Washington improved and expanded during the last twenty years of the nineteenth century, so did the working relationships between the Signal Service and the commercial telegraph corporations. Relationships of suspicion, mistrust, and occasional rancor became respectful, amicable, and supportive. In the 1870s, the Signal Service and Western Union, not without some difficulties, reached greater mutual understanding and accommodation, in part because Western Union continued to consolidate and acquire rival commercial firms, and in part because Western Union successfully secured an exclusive contract for Signal Service weather data.[64] Budget problems for the Signal Corps and Signal Service erupted in the early 1880s. One Signal Corps officer, Henry Howgate, exacerbated their problems. Howgate had popularized Signal Corps involvement in polar expeditions but also had

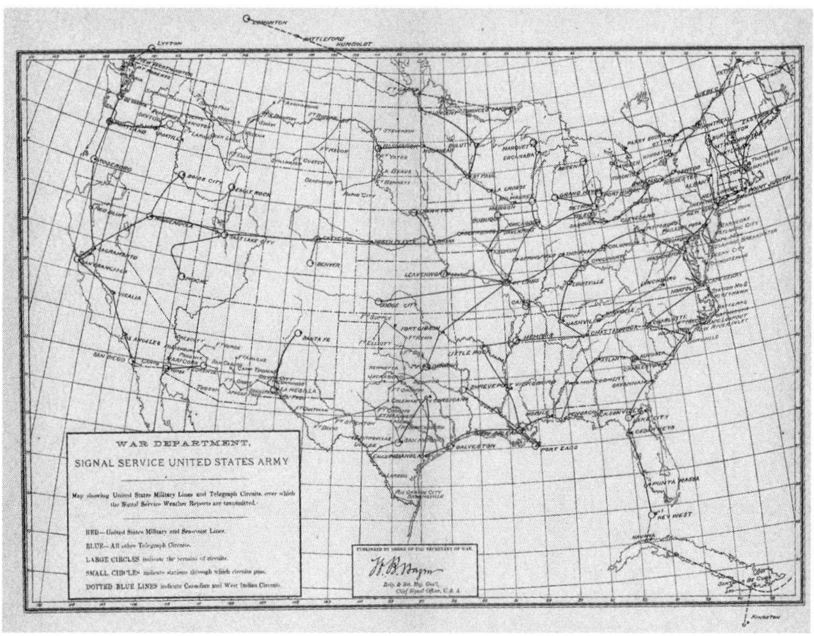

An 1883 map of military telegraph and commercial telegraph network for Signal Service signal flow (*1883 Annual Report of the Chief Signal Officer*, p. 56)

embezzled a significant quantity of Signal Corps funds, leading to several congressional investigations of the Signal Corps in the 1880s. Faced in 1884 with a lack of Signal Service funds to fully pay for telegraph notices of storm warnings to and from some Pacific Coast areas, Chief Signal Officer William Babcock Hazen praised Western Union in his annual report. "The Western Union Telegraph Company, however, came to our relief and actually performed a large amount of service for which it received no pay and for which it declined making charge."[65]

On many occasions in the 1880s and 1890s, Western Union provided access to its networks at no cost for weather data, particularly in times of severe storms or shipwrecks on seacoasts. In addition to the growth in amity, new telegraph technologies doubtless helped strengthen the accord between the Signal Service (and, later, the National Weather Service) and the commercial companies. Duplex, quadruplex, and other multiplex signal flow developments allowed more and more messages to be carried simultaneously on a single wire. More and more relays on telegraph routes were done by print telegraphy rather than by listening and hand-keying.[66] The railroads had built many new lines as they laid track in western North America, and the commercial telegraph companies also had built new lines. Wooden poles were replaced with iron masts, some lines in major urban areas were now routed underground, and any new route or rebuilding of an existing route meant adding more telegraph lines to that route. The telegraph pole was no longer a pole with only one or two strung wires, but a pole with multiple crossbars and a score (or more) of telegraph, telephone, and electric utility lines. Bandwidth had become plentiful. The Signal Service had refined its various codes and ciphers for weather data, and the problems of error resolution for Western Union and other commercial services faded. The growth of the telegraph as a sociotechnical system in the coming decades better served both the military and the corporate. As the nineteenth century closed, the conjoined goals and benefits of the federal government, the military, and the electronic communication corporations progressed on an increasingly harmonious trajectory.

4

Dreams of a Boreal Empire, Nightmares of a Polar Vortex

The Arctic

For practical purposes we may regard the whole telegraph system of the American continent as our own.

—William Henry Seward to Zachariah Chandler, May 14, 1864[1]

The Western Union Telegraph Expedition to Russian America (Alaska), British Columbia, the Bering Strait, and Siberia in 1865–67 explored the possibility of building a telegraph line through the Pacific Northwest and across Alaska. The route planned to traverse the Bering Strait by undersea cable and connect to a telegraph line in Siberia, thus linking North America and Asia (and by extension, Europe) via intercontinental telegraph networks and undersea cables. The expedition included a scientific team organized by the Smithsonian Institution and led by Robert Kennicott. Kennicott was the nation's leading naturalist of the Subarctic and Arctic, having explored those regions in the late 1850s and early 1860s, where he compiled major specimen collections as well as some of the first translation dictionaries of indigenous subarctic languages. A founder of the Chicago Academy of Sciences and the organizer of the first natural history museum at Northwestern University, Kennicott died in Alaska on the Western Union expedition.

Despite the death of Kennicott and the failure of the expedition to establish the telegraph and cable route, the work and research of Kennicott and the scientific team (which included William Healey Dall and Henry Martyn Bannister, both on their first visits to Alaska) was a factor in bringing about the Alaska Purchase of 1867. Although the Alaska Purchase would have likely seen congressional approval without expedition findings, the reports from

Dall on the diversity of Alaskan flora and fauna, along with the meteorological data compiled by Bannister showing that some areas of Alaska had a temperate climate for many months, were among the findings cited by those in Congress who supported the purchase. Rumors of the potential for Alaska to hold significant gold deposits did not hurt those in favor of the purchase, and those rumors eventually proved true in the Klondike region, although the expedition had not extensively explored that area. Having harvested fur-bearing animals to near extinction, the Russian revenues from Alaskan fur and resource extraction were in steady decline. The 1867 purchase was the "mutually beneficial result" of several years of on-and-off negotiations between the United States and the Russian Empire.[2]

Coincident with the 1867 purchase, the military confirmed its permanent presence in Alaska, including the eventual development of electronic communication network facilities. The Alaska Purchase also meant that native Alaskan "Russians" were now native Alaskan "Americans." This brought the indigenous peoples of Alaska into ongoing contact and interaction with the military and, over time, the expansion of government and commerce. The project brought together natural history, the military, indigenous Alaskans, landscapes and ecosystems, and one of the earliest American moments of pursuing global electronic communication networks. This story of Alaska, Kennicott, Dall, Bannister, the Smithsonian Institution, the Chicago Academy of Sciences, Northwestern University, the army, the telegraph, and Western Union reorients the history of the continental-to-global reach of American electronic communication networks toward the Arctic. Reaching toward the Arctic via American electronic communication networks was consonant with other nineteenth-century arctic activities by American entities such as whaling and early navy expeditions. Some politicians and expansionists had begun to imagine the Arctic as a geospatial locale for American expansion and as a landscape and ecosystem for the American pursuit of natural history.

The imagination and pursuit of the Arctic as a locale for American expansion came forth from several individuals and institutions in the first half of the nineteenth century. These individuals and institutions included American whalers, fur traders, and the navy. Some politicians and expansionists looked to the subarctic and arctic regions of the Pacific Northwest as targets for American territorial growth. In addition to individuals and institutions interested in whaling, the fur trade, the reach of the navy, and territorial expansion, the interest of American scientists and natural historians in the Arctic, Subarctic, and Pacific Northwest increased. Spencer Fullerton Baird and others at the Smithsonian Institution looked for further opportunities to study these areas. In the 1850s and 1860s, a number of natural historians and newly emergent institutions in and near Chicago collectively pursued natural history in these northerly regions. Natural history and specimen collections from these regions were one of the major factors leading to the formation of the Chicago Academy of Sciences in 1857. That same year, Northwestern University began to organize its own natural history museum. President of the Northwestern Board of Trustees John Evans and others began the museum initiative two years after first offering classes. Evans and Northwestern turned to a young, talented natural historian named Robert Kennicott, whose family had left New Orleans in 1836 to establish a commercial nursery and horticulture operation near Chicago.

Natural History, Telegraphs and Railroads, and Robert Kennicott

It was the summer of 1855, and young Robert Kennicott rode the rails free as a bird, wherever and whenever he pleased, or at least wherever and whenever the Illinois Central Railroad ran. On that rail line, he had a free pass for unlimited travel. The Illinois Central Railroad sent him in pursuit of nature: to study, learn, teach, collect, and disseminate the natural history of the landscapes and ecosystems the railroad spanned. Already a published scholar, he had studied with or established a continuing correspondence with many leading naturalists of North America, including Increase Lapham, Spencer Baird, Jared Kirtland, Philo Hoy, and Louis Agassiz.[3] Like his father John, he contributed publications to the *Prairie Farmer*, and he probably joined his father in Springfield for the debate between Abraham Lincoln and Stephen Douglas at the 1854 Illinois State Fair, as well as the debate discussion convened at the fair by the Farmer's Assembly on October 3, 1854.[4] A talented American herpetologist, Kennicott carried out some of the earliest studies of the potency of venom from Massasauga rattlesnakes, finding their venom was harmless to animals when ingested in the stomach. He had honored one of his mentors, Jared Kirtland, by naming a previously unknown snake he discovered with Kirtland's name.[5] Kennicott explored Illinois natural history via the Illinois Central Railroad during the summer of 1855 and celebrated his twentieth birthday on November 13 of that year.

Over the next eleven years, Kennicott would cofound the Chicago Academy of Sciences and be appointed as faculty and curator of the natural history museum at Northwestern University. He took up residency at the Smithsonian

Robert Kennicott (Kennicott and Bannister Papers, Northwestern University Archives)

Institution and was a member of that institution's Megatherium Club. Kennicott undertook the first major scientific expedition by an American to the subarctic and arctic regions of western North America and compiled the first translation dictionaries of Native American subarctic and arctic languages. He became the first American naturalist to set foot in and study natural history in the interior of Russian Alaska. Kennicott helped prove (contrary to British maps) that the Yukon River empties into the Bering Sea rather than the Arctic Ocean. He trained a wide range of natural history specimen collectors across the northern regions of the continent. The field collectors included indigenous peoples, who sent countless specimens from subarctic and arctic regions to the Smithsonian. As a Smithsonian curator and its leading specialist in subarctic and arctic regions, Kennicott organized and led the 1865–67 Western Union Telegraph Expedition scientific team. During the expedition, Robert Kennicott died in May 1866 at Nulato, Russian America: hundreds of miles upstream from the Bering Sea and the mouth of the Yukon River.[6]

With little to no formal education beyond home schooling with his siblings, Kennicott began his birding and botanizing at an early age by exploring the landscape and ecosystem around his boyhood home, the Grove, in what is now Glenview, Illinois. Kennicott may at first glance seem a throwback to an earlier age of North American naturalists such as John Bartram and his son William, Constantine Rafinesque, and John James Audubon.[7] In some ways, Kennicott echoed these and other earlier North American naturalist predecessors, like them dedicated to spending an extraordinary amount of time in the field. Kennicott's interests ranged widely, in ornithology, zoology, botany, entomology, and herpetology. He was mentored and tutored by older, more established naturalists, maintaining correspondence with a large network of individuals. Kennicott also was a transition figure, one of the earliest examples of new approaches and methods in North American natural history. From 1855 until his death in 1866, Kennicott was deeply involved at one point or another with three separate corporations, working in conjunction with those corporations in the pursuit of natural history: the Illinois Central Railroad, the Hudson's Bay Company, and Western Union. Kennicott's career involvement with corporations was distinct from that of earlier North American naturalists.

In addition to corporations, Kennicott worked with new and emerging young universities such as Northwestern and with new and emerging young scientific institutions such as the Chicago Academy of Sciences. He was among the first naturalists to work extensively with Spencer Baird at the Smithsonian Institution, where Kennicott took up residency in 1859 (including living quarters) and became a curator.[8] Kennicott remained extraordinarily active with the Smithsonian for the rest of his life and recruited scores of collectors for the Smithsonian throughout the subarctic and arctic regions of western North America. These collectors included officers and workers at various outposts of the Hudson's Bay Company and also indigenous peoples. Kennicott was among the first North American naturalists to extensively work with and train indigenous peoples as collaborators and specimen collectors for the Smithsonian. Kennicott had previously organized a network of specimen collectors as part of his work with the Illinois Central Railroad, writing Philo Hoy in 1855 that he was hired by the railroad "to be that *special agent* to give instructions to Illinois Central R. R. agents about making collections."[9]

Kennicott was among the first North American naturalists to work closely with the emerging new technologies and infrastructures of mid-nineteenth-century North American transportation and communication: the railroad and the telegraph. Kennicott, Baird, and other American naturalists increasingly used the new technologies of transportation and communication to enhance their own fieldwork and observation. The railroad increased the potential range of individual fieldwork across landscapes and ecosystems, and the telegraph augmented ongoing correspondence among naturalists, collectors, and museums. When Baird took up his Smithsonian appointment in 1850, he arrived with two railroad boxcars filled with specimens he had collected in eastern North America.[10] Aside from the enhanced possibilities for fieldwork, the railroad and telegraph also proved crucial in organizing and expanding the scale and scope of specimen collection and the delivery of field collections to the Smithsonian and other new American museums. Both the railroad and the telegraph experienced early expansion in North America coincident with new, organized, systematic efforts at the Smithsonian to collect, classify, and study specimens encompassing the entire scale and scope of North American biota.[11]

Baird (with Smithsonian director Joseph Henry) reached out to recruit potential specimen collectors across military networks, government networks, and corporate networks. One successful corporate effort ensued with the Hudson's Bay Company. The overall networking and recruitment efforts of Henry and Baird quickly paid rich dividends. As William Fitzhugh noted, "Two boxcars was 'peanuts' to the flow of collections that soon began to arrive."[12] Kennicott took the leading role in subarctic and arctic natural history exploration and specimen collection for the Smithsonian, particularly in building and maintaining a network of collectors, many of them associated with the Hudson's Bay Company. With Kennicott largely doing fieldwork in Rupert's Land and the Northwest Territories of British North America (now parts of the western provinces and territories of Canada) from 1859 to 1862, the network of subarctic and arctic collectors Kennicott helped establish at that time are still considered to be among the most prolific collectors the Smithsonian ever enjoyed.

North America, Russia, and Visions of Intercontinental Connectivity

The run-up to the Western Union Telegraph Expedition came about under very different circumstances.[13] An idea first conceived in the 1850s, the concept of an intercontinental telegraph line and undersea cable linking North America and Russia had been reported in newspapers and discussed between Russian and American telegraph interests for about a decade. Various proposals and plans were put forth in the 1850s by American, Russian, French, and British individuals, with eventual success and progress from a plan advanced by Perry McDonough Collins of the United States.[14] At one point, Western Union president Hiram Sibley arranged for a visiting Russian delegation to observe telegraph construction on the Great Plains.[15] The Civil War did not dampen enthusiasm for such a project. In 1862 Senator Milton Latham of California rang out a common theme of telegraph lines and global comity when he proclaimed "We hold the ball of the world in our hand, and wind upon it a network of living and thinking wire, till the whole is held together and bound with the same wishes, projects and interests."[16] Western Union took significant action on the project in 1864. Although the Smithsonian was contacted early on to discuss the possibility of an attached scientific expedition, one important difference from the earlier activities of Kennicott was that the Western Union Telegraph Expedition was not initiated by Baird, Henry, and the Smithsonian. It did not stem directly out of the proven networks of contact and proven methods of success for subarctic and arctic fieldwork and specimen collection previously established by the Smithsonian and Kennicott.

Furthermore, as discussed in earlier chapters, this was 1864, the Civil War was continuing, the United States Military Telegraph (USMT) Company had been formed for the duration of hostilities, and Western Union was intimately involved in the USMT. The Western Union Telegraph Expedition was in some ways a military operation. Kennicott and the scientific team had military appointments in the Union Army, arranged by Western Union officials (including Western Union officials who at that time held military appointments via the USMT). The Western Union Telegraph Expedition even had the faintest trappings of a chartered company, operating beyond the borders of the nation in conjunction with the federal government and the military. In Alaska and Siberia, the expedition members interacted with the Russian-American Company. Unlike the Russian-American Company, however, Hudson's Bay Company, or other chartered companies like the East India Company, there was no pursuit of furs, spices, minerals, or other extractable commodities. The Western Union Telegraph Expedition went in pursuit of signal flow and data, conquering the Arctic to gain the commodity of telegraphic communication between two continents.

There was a blurring here, an ambiguous olio of both a corporation and an army: Western Union and the Union Army to be exact, with just a dash of a chartered company sprinkled in among the main ingredients. In that curious mix, the telegraph was the bonding agent linking the military with the corporate and spreading that relationship beyond the borders of the United States into the far northern reaches of western North America, the Arctic, Russian Alaska, and along the Siberian coast. At a February 1866 session of the House Committee on Foreign Affairs, James Brooks said the project was "not a mere corporation" but "a great act of civilization," and Nathaniel Banks agreed, adding "this enterprise is one of a purely public character."[17] The Atlantic cables of Cyrus Field had yet to prove reliable, having repeatedly ceased to operate soon after being laid. The consistent failure to establish a consistently working Atlantic cable dating back to 1858 may have given rise to a shared perception that a Pacific cable via Russia was necessary for any intercontinental cable and telegraph connectivity between Europe and North America.

This sense of necessity was one important perception driving the interest in and pursuit of a Pacific cable via Russian Alaska and the Bering Strait. There may have also been the specter of British censorship for American and Russian diplomats and other individuals, particularly regarding the geopolitical influence stemming from the future perceived growth of intercontinental cables. Should an Atlantic cable ever prove reliable and then be the only intercontinental cable linking Europe and North America, that might mean potential distortion of Civil War news by British newspapers, and by Reuters and other news agencies. As American Minister to Russia Cassius Clay observed in May 1863 while musing on a future world with both Atlantic and Pacific cables, neither the United States nor Russia was interested in "having the telegraphic intelligence of the world confined to the Atlantic Line, in the sole possession of the British Nation."[18] Thus, a Pacific cable via Alaska and the Bering Strait was not only viewed as a potential necessity for accomplishing a workable intercontinental telegraph and cable infrastructure, but also viewed as an important geopolitical safeguard for the United States and Russia. Cables beneath both the Atlantic and the Pacific oceans connecting North America and Europe would mean no single great European imperial power—whether it be the United Kingdom, Russia, France, or any other rising European empire—would hold an exclusive monopoly on intercontinental telegraph and undersea cable signal flow.

William Henry Seward and the Geopolitics of Electronic Communication

American geopolitical interests were best served by having multiple cable and telegraph systems spanning both oceans, creating multiple metropoles for global communication. In a world of multiple European telegraph and cable metropoles for global communication, potential future conflicts between the United States and any single European power meant the risk of censorship from that power could be minimized by alternative paths for global signal flow, thus enhancing a more secure command for American global telegraph and cable communication. On the North American continent, of course, the geopolitical power of global telegraph and cable communication would, as Secretary of State William Seward noted, be the United States. When Seward observed for "practical purposes we may regard the whole telegraph system of the American continent as our own" he recognized the practical purposes of controlling North American electronic communication networks for the successful pursuit of geopolitics.

The practical purposes of Seward and telegraph-cable geopolitics were prescient for the United States and North America. Of all the established and ascending global imperial and geopolitically inclined powers of the mid-nineteenth century, the United States faced distinct challenges for global telegraph and cable communication. To successfully span both oceans with undersea cables and thus ensure multiple modes of global communication was a huge undertaking. For Seward and the United States, this was not so much a question of telegraph and cable connectivity with the entirety of a planet pursued by the powers in competition for territorial conquest. Rather this was a question about the physical distances between the United States and all the other established, rising, and competing powers. To effectively communicate in an independent uncensored voice via cable and telegraph with the capital cities and major metropoles of the other competing geopolitical powers of the mid-nineteenth century, the United States and only the United States eventually had to successfully span via cable and telegraph the vast distances of two oceans.[19] Regarding "the whole telegraph system of the American continent as our own" was a vital strategic linchpin—a necessity—of Seward's geopolitical vision.

For Seward, there was more at stake for the telegraph and western North America than emergent geopolitics. He saw the Alaska telegraph project and the Pacific cable as central to global trade and amity, something that would end "isolation" for both the American and the European continents. Seward forecast that "if our telegraph system should be connected by a transoceanic line with the *one* that is now performing its equally isolated part on the eastern continent, new and stronger reciprocity would be exhibited." He recognized that "each of the isolated telegraph systems existing on the two continents is imperfect and incomplete." A transoceanic telegraph and cable system across the Pacific had important domestic ramifications for the United States and its Trans-Mississippi states and territories. Telegraphy growing and expanding across the continental expanse of the nation would "preserve peace and order at home" and "maintain a military and naval police" on inland frontiers and seashores. Beyond the cultivation of peace was the greater promise of utilizing an expanding American telegraph system to cultivate civilization. This goal needed to include the interior of western North America, not just eastern North America and the states and territories of the Pacific coast. Seward believed this was a duty of governance and citizenship: "it seems to me that in extending dominion over inland mountain regions, and setting up the national flag on the Pacific coast, the American people, however inconsiderately, assumed the duty of diffusing an equal civilization throughout the whole of the great country which they have thus included within their borders." Concluding his analysis with a view toward the future, Seward opined that "the great interests of human society will, at a very early period, require more than one, and more than two, trans-oceanic, world-encircling telegraphs."[20] The idea of connecting North America and Russia by telegraph, a project called the "Russian Extension," took root.

The Smithsonian, the USMT, and Western Union

In April 1864 Robert Kennicott wrote a long letter to his colleague Roderick McFarlane at the Hudson's Bay Company post Fort Good Hope in which he included discussion of the Western Union Telegraph Expedition plans underway, despite the Civil War.[21] Kennicott surmised McFarlane likely had one-sided information about the war. "I suppose you read no end of croaking over our supposed downfall in the English papers. We are carrying on a horrible war, but it is necessary." Kennicott confirmed it "is now finally determined there shall be no slavery in the United States and that the government never yields an inch of its territory to another. When this war is ended we will probably have one with France about her occupation of Mexico." Kennicott then went on to boast of new American projects. "If you read yarns about our supposed downfall let me call your attention to our daily increasing prosperity and the gigantic projects about here." He told McFarlane about plans for the transcontinental railroad and canal projects linking the Great Lakes and the Mississippi River. He also broke the news that "some enterprising Yankees have a grant from the Russian and British governments to make a telegraph line from European Russia by Bhering Straits and through British Columbia to the United States." Kennicott had met with these individuals but claimed to have been discreet. "The parties interested came to me for information. But as I don't know that the Hudson Bay Co. cares to have such an exploration of their territory made I was 'extremely ignorant of the country' and shall continue to be so unless I learn that the company wishes it to go through."[22] Kennicott then shifted the topic from nation to nature, finishing the letter with several pages of advice about specimens, collecting, trapping, skinning, preserving, shipping, Great Slave Lake, and Great Bear Lake. Kennicott closed by reminding McFarlane that "Indians could collect lots of shells" and assist the Smithsonian in building up malacology specimen collections.

Kennicott may have written McFarlane for another reason. The planned route for the telegraph to Russian America and across the Bering Strait was not decided. One proposed route began at St. Paul, Chicago, or Detroit and proceeded northwest along the Red River (of the North) to Lake Winnipeg and through the Mackenzie River watershed (including tributaries such as the Liard River) to "Peel's" and then "across the Rocky Mountains to La Pierre's House" and "down the Porcupine (River) to the Youkon." Much of this route could utilize "the regular boat navigation of the Hudson's Bay Company" in river travel, lake travel, and portages. Such a route would traverse the Rocky Mountains at the northern limits of that mountain range and might also consider running the route from the mouth of the Mackenzie River along the coast of the Arctic Ocean across Alaska to the Bering Strait.[23] William Pullen, a British navy officer, had successfully navigated the Arctic Ocean coastline from the Bering Strait to the mouth of the Mackenzie River in 1849. Other advantages to this northwesterly route running largely through Rupert's Land and other British possessions in subarctic areas north of the United States included known transportation routes through a navigable lake and riverine system, pasturage and forage for draft animals, reliable game and fish, and a decent supply of timber. Furthermore, the route utilized an existing network of Hudson's Bay Company posts, houses, and forts. Finally, the route was ideal for a growing project: collecting specimens from subarctic and arctic regions for the Smithsonian Institution.

The document discussing this telegraph route has no names indicating who participated in the conversation. The route was undoubtedly suggested by Robert Kennicott. The route traversed known territory, utilized known transportation routes, and utilized Hudson's Bay personnel. The route encompassed landscapes and ecosystems explored by Kennicott in the late 1850s and early 1860s. Such a route would lead to greater success and coordination of subarctic and arctic specimen collection for the Smithsonian and, from an environmental perspective, the route was known to have sufficient timber, pasturage, game and fish, and water-based transportation for equipment and personnel. A route utilizing the Hudson's Bay system also brought something else: the possibility of collecting meteorological data by telegraph. It is likely that Joseph Henry would have approved of collecting weather data, for he had already begun experimenting with weather forecasts that were based on telegraph reports. Synoptic meteorology could have emerged across the Kennicott telegraph route. Had that happened, data records of North American climatology would be different. Weather data from a large area of the subarctic and arctic might have been used for North American synoptic meteorology as early as 1870, along with the data collection of the Signal Service. Despite its careful consideration of personnel, transportation, climates, and ecosystems, the route was not to be.

The Transcontinental Telegraph and the Russian Extension

The route chosen for the telegraph line across western North America to the Pacific coast and then onward to Russian Alaska, the Bering Strait, and into Asia was instead a route that began with the Transcontinental Telegraph from Chicago to San Francisco, and then north along the Pacific coast into British Columbia. Western Union in part settled on this route for financial and corporate reasons. The route would utilize existing telegraph lines and companies already owned by Western Union or lines Western Union expected to soon control as part of its "gobbling up process of over half the California State lines." This included the recent purchase of "a good new line recently built way up to Vancouver Island" that would be a "good feature for the Russian Extension" and also on its own "pay favorably" in Pacific coast telegraph revenue.[24]

Revenue was important for Western Union, but something else was even more important: security. Using the Transcontinental Telegraph as the first leg of the route for the Russian Extension also meant that the Transcontinental Telegraph would be under the continual protection of the Union Army from attacks by Great Plains warriors. In 1864 a congressional act "to encourage and facilitate Telegraphic Communication between the Eastern and Western Continents" authorized the secretary of the navy to detail a vessel to assist in surveys and soundings for the expedition. Branch or extension lines to "various mining districts and other settlements" in Trans-Mississippi territories were encouraged and "the right to take timber and stone for construction purposes" from public land was codified. A land grant of "forty acres for each fifteen miles of line constructed" and a "permanent right of way over any unappropriated lands of the United States" were offered. The act encouraged military interest, calling for the right of the federal government to "connect said line or lines by telegraph with any military posts of the United States" and to use the line for government purposes. Furthermore, the act ensured military protection and security. The secretary of war was authorized by the act to instruct military officers to "use any available force at their command" to protect the telegraph lines, including protection "from injury by savages or other ill-disposed persons."[25]

In September 1864 Russian Minister to the United States Eduard De Stoeckl wrote Russian Minister of Foreign Affairs Alexander Gorchakov with a progress report on planning the Russian Extension. Assuring Gorchakov of good progress, Stoeckl also had visited New York to "collect some background information" and meet with Hiram Sibley. He found Sibley to be "an enterprising and energetic man possessed of a large fortune" and someone who had already proven by his success with the Transcontinental Telegraph that he could build a telegraph route "despite seemingly impossible odds." Stoeckl was convinced it would be "easy to construct and maintain the telegraph line in well-settled areas" but would "prove problematic where the country is only populated by savages." Sibley believed "friendly relations with the Indians" could be established through gift-giving. Stoeckl confirmed the "same obstacles were present during the construction of the San Francisco line," which ran through "deserts solely populated by Indians." Initial hostility was now being met with "gradual success at establishing good relations" with indigenous peoples. Offering a variation of an emergent theme, Stoeckl surmised the growing success of good relations could be in part because the Indians "attribute a mysterious energy to the telegraph, calling it 'the work of the Great Spirit.'" But Stoeckl found one thing "odd" about the attitudes and actions of indigenous peoples toward the Transcontinental Telegraph. He noted "though they are presently at war with the Americans and attack and loot homes on the frontier, they have not caused any damage to the telegraph system that crosses these same areas of habitation, and we continue to receive regular dispatches from California."[26] Stoeckl was right. It was odd that, in comparison with other attacks and conflicts on the frontier, the Transcontinental Telegraph had not yet been a zone of conflict. That would change dramatically, however, during the next three months. Any lingering doubts about the value of military protection vanished in November 1864 after the Sand Creek Massacre and its 1865 Julesburg aftermath.

The route was established and plans proceeded, now with the beneficial security of military protection. At an October 1865 meeting for Western Union stockholders, the board of directors reviewed the progress on the Russian

Extension. They acknowledged that the Transcontinental Telegraph had, since the start of the year, been "seriously interrupted" in the Julesburg area "by Indian raids and depredations, materially diminishing the receipts" from that route. Security and revenue of the Transcontinental Telegraph were high priorities. The directors added that Western Union had "now but a single wire from Omaha to Salt Lake" via Julesburg and "what is absolutely required to afford the needed facilities" for "our business with the Pacific Coast" was a new line from Denver to Salt Lake City that bypassed Julesburg. The expansion would result in more wires, including "one of them as far south of the region of Indian depredations as to be substantially free from disaster" and a "second wire from Julesburg to Omaha, on the poles now in use" which would increase capacity and also be important "for the Russian line."[27] The Kennicott route was environmentally sustainable but held few prospects for significant new revenues from the territories which it would traverse. The utilization of newly acquired California and Pacific coast lines on the Russian Extension route was financially profitable. But beyond all else, the Russian Extension was protected by an act of Congress that legitimated military protection of the Russian Extension. Therefore, a Russian Extension route to the Bering Strait that included the Transcontinental Telegraph from Chicago to San Francisco brought a congressional mandate for military protection and security to the Transcontinental Telegraph. The Transcontinental Telegraph and the Russian Extension were woven together. A congressionally mandated security blanket now swaddled the Russian Extension and, perhaps more important, also swaddled the Transcontinental Telegraph.

In 1864 Secretary of the Treasury Salmon Chase met with Hiram Sibley of Western Union to discuss the Russian Extension. Sibley agreed that military posts and the government "shall at all times be entitled to and have priority in the use" of all Russian Extension lines as well as the privilege to connect any existing or new military post to the line. Sibley also agreed to allow the line to be used "at any time free of charge, to the Coast Survey, the Smithsonian Institute, and the National Observatory, for scientific purposes." Including the Coast Survey helped strengthen the commitment from the navy, which had been at issue. A navy officer had earlier balked at the idea of the navy using "a Man of War for the transportation of telegraph material." Western Union officials knew this officer was on "intimate terms" with Cyrus Field and wondered whether this particular officer was indicative of a drift in the Navy Department to favor Field and his Atlantic cable efforts over Western Union and their Russian Extension efforts.[28] The navy did eventually provide Navy ships in support of Sibley and the Russian Extension. With the promise of support to the Coast Survey, the Naval Observatory, and the Smithsonian, Western Union became a patron of American science in exchange for promising free transmission of a smattering of telegrams.[29]

The Arctic in Chicago and the Midwest: Museum Collections

From the earliest involvements of Baird and Kennicott with the Western Union Telegraph Expedition, all parties intended to procure natural history specimens for both the Smithsonian and the Chicago Academy of Sciences. In March 1865, a memo of understanding shared expenses and specimens between the two museums, with Kennicott committing one thousand dollars in Chicago Academy of Sciences funding.[30] This cost sharing and specimen sharing was not unprecedented. Kennicott launched the Chicago Academy of Sciences in part because of a shared vision with Baird and the Smithsonian that the Chicago Academy would, as Kennicott told McFarlane, become "a sort of *Young Smithsonian*" with extensive subarctic and arctic collections.[31] Kennicott had embarked on his earlier subarctic and arctic research with an understanding that Baird and the Smithsonian would be open to donating duplicate specimens collected by Kennicott and his network to the Chicago Academy. The donated duplicates from these missions boosted the stature of the Chicago Academy, securing an Illinois state charter and raising sixty thousand dollars in funding with the encouragement of Kennicott and Louis Agassiz, who both lectured in Chicago about the value of the duplicate specimens in February 1864.[32] Some duplicates from these Kennicott expeditions also made their way to the natural history museum at Northwestern. The 1863–64 Northwestern University Catalog of Study told students the museum held more than 2,500 specimens. "The collections have been labelled by experts in each department, under the auspices of the Smithsonian Institution, and are believed to possess unusual advantages for investigation and study."[33] Oliver Marcy arrived at Northwestern in 1862 and eventually replaced Kennicott as the director of the Northwestern science museum. Marcy was professor of physics and natural history and by 1870 was appointed as museum curator. Before the Western Union Telegraph Expedition set off for Alaska, Marcy visited San Francisco in 1865 and met with Dall (who himself was involved with the Chicago Academy before the Western Union Telegraph Expedition). Marcy and Dall saw each other when Marcy was hired as a geologist for work on a proposed road between the mouth of the Columbia River and Montana Territory.[34]

The organization of the scientific team for the expedition began in Chicago. Kennicott made his initial plans, and recruited some team members, by working with the Chicago Academy of Sciences (William Dall) and Northwestern University (Henry Martyn Bannister). Kennicott assembled the Chicago members on visits and also while in residence at the Smithsonian. When Kennicott first joined the Smithsonian in 1858, John Evans wrote Kennicott a letter of introduction to Illinois senator Steven Douglas, telling Douglas that Kennicott was the curator of the Northwestern museum.[35] Evans hired Kennicott as the museum curator, knew the Kennicott family, and stayed at the Grove on several occasions. Other members of the Chicago Academy offered advice on the expedition, as did individuals at Northwestern. The team traveled to Washington and in Washington recruited others for the scientific expedition. They made their way to New York to begin their voyage in March 1865. Then it was on to San Francisco, the operational headquarters for the Western Union Telegraph Expedition. Kennicott, Bannister, Dall, and the rest of the scientific team made the long journey down the Atlantic seaboard, across Central America, and up the Pacific Coast. They sailed into San Francisco in April 1865 and found chaos and turmoil throughout the logistics and preparations for the expedition.

From San Francisco to Sitka

After a miserably organized start to the expedition, poorly equipped, endlessly botched, and constantly delayed in San Francisco by the officers Western Union had placed in charge, Kennicott wrote Baird on July 23, 1865, while anchored just off Vancouver Island on the way to Sitka. Kennicott made his sentiments about the Western Union Telegraph Expedition perfectly clear: "D—n all corporations!"[36] He may well have added damnations to the army, the USMT, and Edwin Stanton. The entanglements of the Western Union Telegraph Expedition with the nexus of the telegraph and the Civil War lay at the roots of a huge range of problems that would vex—even unto death—the scientific corps of the expedition. When William Dall later wrote Spencer Baird in September 1866 about the death of Kennicott, Dall laid the onus for the difficulties, mismanagement, and disasters of the scientific component of the expedition—and even the untimely death of Kennicott—squarely at the feet of expedition commander Colonel Charles Bulkley and his military command staff. "There is no doubt he was badgered to death [by Bulkley and other officers] . . . no one but me knows the fiery furnace he worked through in '65 in San Francisco. A legion of hell devils baiting him all the time."[37] The leader of the hell devil legion came from Western Union and the USMT.

Charles Bulkley (Dall-Healey family photographs, William Healey Dall carte de visite album, ca. 1858–1911, Charles S. Bulkley, ca. 1861–65, photo 69.11; Collection of the Massachusetts Historical Society)

Colonel Charles Bulkley commanded the Western Union Telegraph Expedition. Bulkley was an army officer by way of the USMT, among the individuals (along with Anson Stager and Thomas Eckert) appointed as Union Army officers by Edwin Stanton. Bulkley had extensive telegraph construction experience prior to the war, having played a central role in constructing telegraph lines in 1847–48 between Washington and New Orleans, lines constructed in part as a reaction to the lack of telegraph connectivity during the war with Mexico.[38] As a USMT officer, Bulkley spent much of his post during the Civil War in the Department of the Gulf, where he had received by Stanton's orders a significant amount of telegraph equipment originally procured for Albert Myer and the Signal Corps.[39] Under the command of General Nathaniel Banks, in April 1864 Bulkley built a military telegraph line from Brazos to Brownsville along the Texas gulf coast. In August 1864 Bulkley was relieved of his duties in the Department of the Gulf at his own request in order to assume the leadership of the Western Union Telegraph Expedition. Bulkley was furloughed indefinitely—not discharged or decommissioned from his military appointment—and took along from his ranks in the Department of the Gulf a handful of additional military telegraph experts.[40]

In October 1864, Bulkley was corresponding with Spencer Baird on USMT letterhead about the upcoming Western Union Telegraph Expedition.[41] Western Union also arranged at some point in 1864 for photographs of Bulkley and Frank Pope (who surveyed routes for expedition telegraph lines in British Columbia) to circulate among some indigenous peoples and communities near Fort Yukon.[42] In September 1864, Secretary of the Navy Gideon Welles authorized a naval vessel to support Bulkley on the telegraph project.[43] As arrangements for a Smithsonian scientific team took shape, Bulkley prepared to depart for San Francisco, telegraphing Baird on December 16 that Baird could "engage half a dozen" scientists if he wished.[44] Western Union directors had ordered that Bulkley be "called on to take command of this expedition."[45] This suggested Western Union, through the aegis of the USMT, exercised a measure of influence in transferring standing officers of the Union Army away from one theater of war and into other Western Union projects, all the while preserving the military appointment of those transferees. Optimism was the mood of the day as press reports and news stories emerged, with *The Telegrapher* predicting in November 1864 "we may confidently expect to be holding a mysterious conversation with the Czar of all the Russias by the summer of '66."[46] The *New York Examiner and Observer* told its readers in 1865 that the project meant "the Arctic wastes will become the medium of the world's intercourse."[47]

The disorganization and endless delays threatened the ability to arrive in Alaska sufficiently before the onset of winter. On May 23 Kennicott also telegraphed Joseph Henry at the Smithsonian, asking Henry to speak to the secretary of the treasury about assigning Charles Melville Scammon, a Revenue Service officer, to the expedition. Scammon, an experienced whaler, was knowledgeable of North Pacific and arctic seas. He was the brother of Jonathan Young Scammon, a board member of the Chicago Academy of Sciences. The secretary of war granted leave to Scammon in order to allow Scammon to join the expedition as one of the ship captains.[48] Kennicott wrote Baird on June 4 that "Bulkley is the executive officer of a big—*very big*—corporation, and he makes a good one." Kennicott lamented the late start: He assured Baird "I told the Directors in New York that the line from the Straits to the head of the Frazer River could be *completed* by the last of October 1866—I said distinctly that if I could commence at the mouth of the Youkon June 1st 1865 I could build it—and I could have done so."[49] A month later and the expedition ships still had not sailed for Alaska. Kennicott wrote Baird on July 9 that since arriving in April he had been "in a perfect *purgatory*" until an announcement today "that we sail tomorrow. I am to go up the Youkon. I must make most of my distance in winter. I may have a tough time."[50]

Science and Engineering in the Field: The Telegraph and Natural History

The Western Union Telegraph Expedition explored territory, surveyed potential routes for telegraph lines, and partially constructed telegraph lines in three areas: British Columbia, Russian America (Alaska), and Siberia. The work in British Columbia began at the U.S.-Canadian border and proceeded northward, relatively near but not along the Pacific coast, sometimes more or less following the path of the Fraser River. In Siberia, routes were surveyed from the Chukchi Peninsula on the Bering Strait going inland roughly along the northern border of the Okhotsk Sea to the mouth of the Amur River and the nearby settlement at Nikolayevsk. In Alaska surveying and exploration took place from the Stikine River in British Columbia northwest via Fort Selkirk and Fort Yukon to the Seward Peninsula and Bering Strait, with much of the route following the Yukon River. This work also entailed significant exploration by the scientific team in and around Norton Sound, including St. Michael (then known as St. Michael Redoubt), Unalakleet, and Nulato. St. Michael Redoubt was an important Russian fortification and strategic deepwater seaport with access to the Pacific Ocean, Yukon River, and the Bering Strait. Sitka, located on the southeast coast, was another important location, although no survey work for telegraph lines was done in Sitka. Another strategic deepwater port with excellent access to the Pacific Ocean and known as the Paris of the North, Sitka was the administrative center of Russian America and the location of the main field offices of the Russian-American Company.

Organizational problems and a military management style continued to frustrate and hamstring members of the scientific expedition upon arrival in

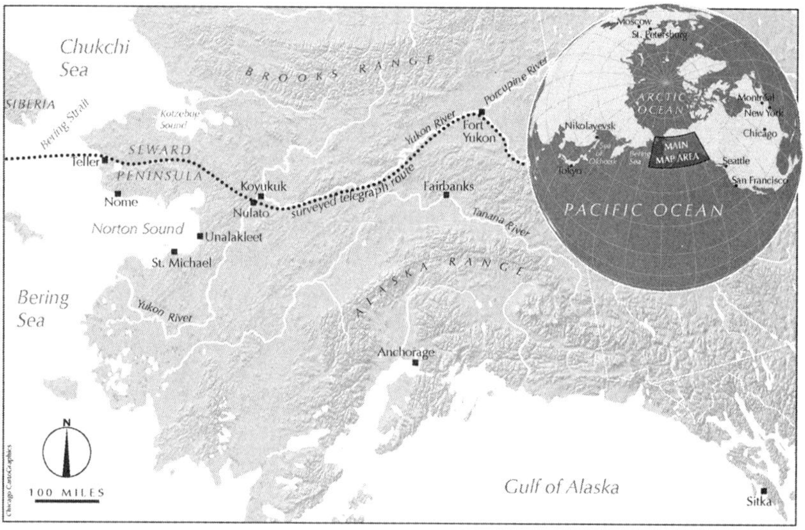

Map of Alaska, including surveyed telegraph route (map by Chicago CartoGraphics)

Sitka in August 1865 and in subsequent expedition departures across Alaska.[51] The scientific expedition members now had military appointments. Kennicott held the rank of major, and Bannister and Dall the rank of second lieutenant. For Dall (and presumably the other military officers from the scientific team) the military appointment came directly from his employment contract with Western Union. The contract of July 1, 1865, between Dall and Western Union for one year of employment stated that Dall entered "into the service and employment of the Western Union Telegraph Company" in the "capacity of Second Lieutenant."[52] The contract may indicate that Western Union, with the USMT still operative, had at that time the authority to commission Union Army military officers during the lingering exigencies of the Civil War. The commission of Dall, taking place after the end of battlefield conflicts, may seem an anomaly nevertheless. It is unclear, however, what, if any, details Western Union shared with Edwin Stanton and the War Department about the USMT and the Western Union Telegraph Expedition. In his final annual report of the USMT of June 1866, Anson Stager wrote Edwin Stanton that as of June 30 "the only officers remaining in the United States service connected with the Military Telegraph" were himself and Thomas Eckert, stationed at the War Department.[53] The statement may or may not have been disingenuous, but it was nevertheless accurate. Buckley, Kennicott, Dall, Bannister, and the other USMT officers on the Western Union

Telegraph Expedition were active in British Columbia, Russian Alaska, and Siberia—not in "United States" service. Alaska was still Russian. Bulkley reported in July 1865 to Western Union that his "organization is military in character, requiring officers and men both in the land and marine service to wear uniforms, without cost to the company" and that their accounting system was "similar to the Quartermaster's Department of the Army." He suggested that a delay in starting off from San Francisco may have been a "blessing in disguise" because the Confederate raider CSS *Shenandoah* had recently been marauding in North Pacific waters.[54] Dall later wrote his father it was known the *Shenandoah* was likely in the North Pacific, surmised the *Shenandoah* did not know of (and may not honor) the recent truce, and that the two cannon on the expedition flagship would have been "no match for the heavy long range rifle of the pirate."[55]

Examples from Alaska and the expedition such as quartermasters, uniforms, military commissions, and Confederate warships are some of the lingering signs of the ambiguous mix of both a corporation and an army.[56] Kennicott told Baird that Bulkley did not appreciate the challenges of Arctic conditions and mistakenly expected men "to subsist on the Army ration and work with Army materials," adding that for Alaska, Bulkley "runs Army a little too far for such a country."[57] Brought together hurriedly by the hand of Edwin Stanton in the crucible of rebellion and an existential crisis, Western Union and the Union Army did not seem at all hurried to dissolve many of the joint attributes and qualities of their own peculiar institution as soon as the battlefield confrontations of the Civil War halted. Anson Stager's 1866 budget report hinted at one component of their lingering intimacy. Bulkley and his military organization, his command structure, and his hand-wringing about marauding Confederate ships in the North Pacific also linger beyond the Civil War.

In some ways, the Western Union Telegraph Expedition and the USMT–Western Union–Union Army commingling raise the specter of the expedition as a sort of occupying army—or better, preoccupying army—in Russian Alaska. Ostensibly preoccupied with building an intercontinental telegraph, USMT–Western Union military officers and men, in some capacity, were continuously on the ground somewhere in Russian Alaska from their arrival in August 1865 until the Alaska Purchase and the subsequent transfer at Sitka on October 18, 1867. Spanning two years and two months, it was the longest continuous overseas presence of any American military force to that date—that is, if it was an American military force. Observing the transfer ceremonies atop Castle Hill in Sitka, Admiral George F. Emmons, commander of the USS *Ossipee* anchored in Sitka Harbor, observed that he was no longer in Russian America. Looking at the American flag now flying alone above Castle Hill, Emmons hoped that flag "will hereafter be the emblem that will embrace all of North America."[58] Military officers, telegraph experts, construction crews, corporate representatives, and a few Smithsonian–Chicago Academy–Northwestern naturalists on a scientific expedition—all these identities were in the mix, all representing the growing power and influence of the United States over western North America, all of them preoccupying Alaska.

Military Mindsets, Naturalists, and Differing Masculinities

Dall did not find much, if anything, useful in holding an officer rank while conducting business with the other officers of the expedition involved in construction and logistics. Kennicott had earlier told Baird that he found the military officers difficult, particularly a lieutenant colonel named William Hyde: "a low lived cunning customer—very Pecksniffian and an errant humbug."[59] Dall wrote his father a long letter in September 1865. He said the expedition was "organized on a military basis, which as it is carried out is a humbug." Dall added "the red tapes and the forms are vigorously pursued, to the detriment of the organization and the interests of the Company." The scientific expedition had military uniforms, but Dall succeeded in getting the "laugh" on "a red headed, arrogant, ungentlemanly ass" (Hyde) who tried to design insignia. Dall "got a special order allowing me action on it, & now wear as Director of the S.C.T. operations shoulder straps with two silver scallop shells on each and a little gold bug in my cap." Having earlier seen Kennicott depart for a Yukon exploration, Dall also told his father "I am now left in charge of the Natural History operations." Dall believed Bulkley had "a lust for power more than ambition" among other troubling qualities. Dall closed by sending his love and reminded his father to try to find more shells, particularly from the Philippines and "from your sea captain friends."[60] Kennicott issued the directive appointing Dall as head of the scientific team for the duration of his absence on July 13, when the scientific team was at sea about thirty miles north of San Francisco, near Drakes Bay.[61] Although such an early order by Kennicott may seem precipitous, it could have been a routine exercise of caution. The voyage north from San Francisco to Sitka could find rough weather and difficult navigation. Kennicott also may have foreseen a growing need for Dall to take command as it became clear that

Kennicott would be taking a small party and immediately head for the interior of Alaska as soon as possible. Kennicott wrote Caroline Wells Healey Dall (Dall's mother) about his decision and his trust in Dall, adding "You have reason to be proud of such a son."[62]

Kennicott had a close relationship with Dall, as well as with Henry Martyn Bannister. He worked with Dall at the Chicago Academy of Sciences and with Bannister at the Northwestern museum when Bannister was a Northwestern student. Both Dall and Bannister were, in a way, to Kennicott as Kennicott had been to naturalists such as Kirtland, Hoy, or Lapham. Both young men deeply admired Kennicott, yet there is scant evidence to suggest Dall and Bannister grew close to each other. Bannister had previously worked with Kennicott at the Smithsonian, specializing in fossil collections. Dall wondered about Bannister in some of his letters, including questions about Bannister's stamina. One day in summer 1865 Bannister found himself participating on a construction team felling trees in British Columbia. Bannister wrote in his journal that "the work which I undertook, peeling and trimming, was not very hard" but "after supper I found myself rather tired and retired for the night in good season."[63] Bannister may have overstated his participation in his one and only day as a pole-shaper. Dall wrote Spencer Baird at the Smithsonian that Bannister had arrived at Sitka from British Columbia and was sent (to Sitka) "as they feared he would not stand the work."[64] Dall also noted in his dairy that Bannister had bouts of seasickness during the March 1865 expedition voyage from New York to Nicaragua.[65] Bannister probably should not have been on a construction team in British Columbia in the first place. Bulkley split apart the Smithsonian scientific team. When Bulkley wrote Kennicott on September 8 appointing Kennicott as "Chief of Exploration in Russian America," Bulkley named the individuals in the party to work directly with Kennicott, a list of names that did not include Bannister or Dall.[66] In a letter to his parents written on July 26, Henry described his British Columbia experiences and discreetly suggested Bulkley, not Kennicott, placed him on the British Columbia construction team. "I stayed a few days with a construction party, cutting poles, and then was sent down. . . . I asked [Bulkley] to be allowed to accompany the expedition up the coast [to Alaska]. . . . I expect [this boat on which I now sail] is to be my home for the season unless I receive further orders to go somewhere else."[67] The scientific team did not have complete autonomy in determining its own locales or working groups. That power was not with the naturalists and the Smithsonian but with the military officers, exercised by Western Union and the USMT command structure of the expedition.[68]

Henry Martyn Bannister (Dall-Healey family photographs, William Healey Dall carte de visite album, ca. 1858–1911, Henry Martyn Bannister [1844–1920], 1865, photo 69.49; Collection of the Massachusetts Historical Society)

Bannister's spyglass from Alaska (Kennicott and Bannister Papers, Northwestern University Archives; author photo)

There was a rivalry of sorts between Bannister and Dall. Bannister was twenty years old and Dall nineteen years old when they and Kennicott sailed out of New York City at the start of the expedition.[69] Both were the sons of religion scholars. Dall's father, a Unitarian minister, was an expert on Hinduism and served as a missionary to India. Bannister's father, a Methodist minister also named Henry, was a professor of Greek, Hebrew, and sacred literature at Northwestern and the Garrett Biblical Institute. Both young men were brilliant in different ways, and both looked to Kennicott as a mentor and leader. Dall thought Bannister was "an awful queer stick and as obstinate as Balaam's mule."[70] Dall knew, and on occasion met with, Bannister's father when Dall was in living in Chicago. The elder Bannister visited Dall at the Chicago Academy on February 15, 1865, and talked with Dall "about Henry."[71] Although Dall apparently never held any formal appointment or position at Northwestern University, his diary indicates he visited Evanston and the Grove on several occasions. The Kennicotts were well known in Evanston, and several members of the extended Kennicott family attended Northwestern. Bannister said in a 1909 interview that "Robert Kennicott while not exactly an Evanstonian, geographically, was pretty nearly one."[72] Dall also wrote father Henry about young Henry on behalf of Kennicott late in 1865 while Dall wintered in San Francisco, telling him that Kennicott "was displeased at H's obstinacy and whims but is now OK."[73] Dall's relationship with his own father was overwhelmingly through written correspondence, for his father had largely remained in India from the time Dall was ten years old.

Dall was born in Boston. Upon graduation from the Latin School in 1863 he bypassed formal matriculation at Harvard and instead began specialized studies with Louis Agassiz, who encouraged Dall's growing interest in shells and malacology.[74] Bannister moved to Evanston when his father took a faculty position at the Garrett Biblical Institute, colocated with Northwestern University, and met Kennicott while a student at Northwestern.[75] Prior to the Western Union Telegraph Expedition, Bannister had shown an interest in local weather conditions, often sharing his findings with Josiah Willard.[76] Willard was the father of Frances Willard, who would become the first dean of women at Northwestern, an early suffragist, and a founder of the Women's Christian Temperance Union. Henry's sister Mary later married Frances's brother Oliver. Both Dall and Bannister grew up with feminists. Dall's mother, Caroline Healey Dall, was a feminist writer advocating the rights of women in public life and was active in the Boston women's rights movement. Bannister grew up alongside the Willard family. His sister Mary later edited the newsletter of the Women's Christian Temperance Union.

William Dall (Dall-Healey family photographs, Healey-Dall family carte de visite album, ca. 1857–81, William Healey Dall, ca. 1861–65, photo 69.214; Collection of the Massachusetts Historical Society)

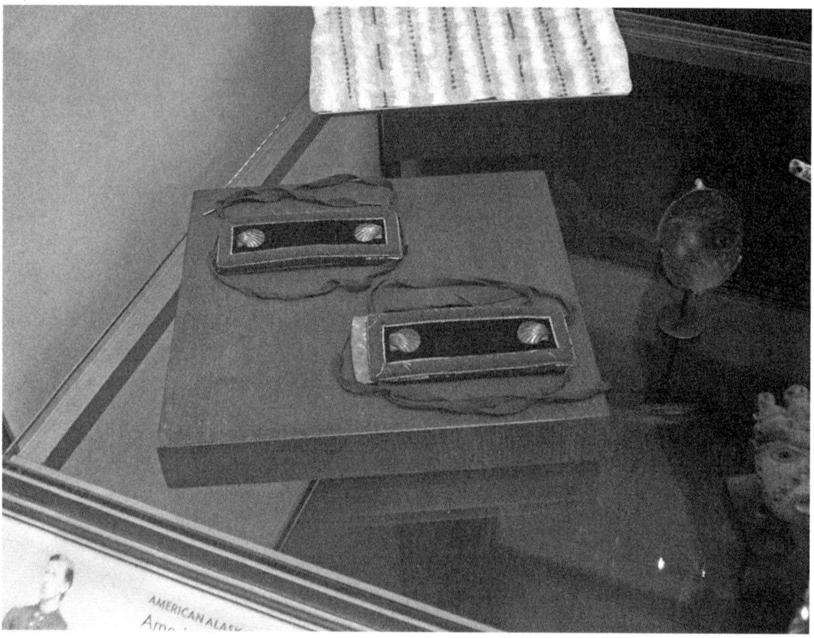

Dall's shoulder straps (Alaska State Library, Archives, and Museum; author photo)

Environmental Masculinity: "The Score of Muscular Accomplishments"

With the onset of their first Alaskan winter in December 1865, Bannister was severely frostbitten while traveling in a raging storm, at one point given up for lost. Bannister in his later years increasingly suffered from physical disabilities in part brought on by this experience. The 1916 *Northwestern Bulletin*, recounting students from the 1860s, told readers that Bannister was "nearly frozen to death in Alaska, having been saved by an Indian, who went back for him after he had been left to die."[77] This was no more than a month after Kennicott and Bannister had shown up at an encampment between St. Michael Redoubt and Nulato in haggard condition from an excursion, arriving "with one Indian, four dogs and a sleigh. Kennicott and Bannister were both lame. Had a very rough trip of it."[78] As a result of the near-death experience in the storm, Kennicott decided to no longer allow Bannister to continue in fieldwork. He stationed Bannister at St. Michael Redoubt. Bannister managed stores and provisions and kept a detailed meteorological journal. Kennicott told Bannister "I beg of you Bannister to be less sensitive on the score of muscular accomplishments. And on no account to feel there can the slightest disgrace attach to such a misfortune as this—An Indian dog can stand more cold, starvation, or fatigue or run faster than any of us, so for God's sake don't brood any more over any failure on your part to endure all you expected to."[79] William Ennis, another member of the scientific team then with the Kennicott party, wrote Henry to "condole with Kennicott and all our boys on account of the misfortune that has happened you." Ennis added Kennicott was torn between duty and returning to see Bannister, noting all could "tell the worry of his mind ever since the intelligence of your sickness arrived." He finished by encouraging Bannister that "although you caught it in this pop, the next time you will be up to *Snuff*" for future activities.[80] Perhaps Kennicott knew of Dall's earlier comments questioning Bannister's stamina, and perhaps not. Perhaps Kennicott suspected that some of Henry's qualms about his own "score of muscular accomplishments" stemmed in part from earlier incidents in Evanston and at Northwestern University.

Kennicott's concern for Bannister's health and safety is understandable. In advising Bannister that there was not a question of duty, courage, stamina, or character, Kennicott may also have been thinking of an earlier tragedy that had a profound impact on the Bannister family: the Lake Michigan sinking of the *Lady Elgin* in September 1860, near the Northwestern campus. Rammed at night by another ship, the *Lady Elgin* broke up quickly as a storm gathered and strengthened. Survivors struggled against the storm and

A RUSSIAN POST AT NULATO

Nulato (*Illustrated London News*, 1868, Alaska State Library, Archives, and Museum)

pounding surf to reach landfall near Winnetka, and a general alarm went out along the shore. Northwestern students responded and rushed north to Winnetka to try to save passengers. Edward Spencer, a Northwestern Garrett Biblical Institute student, was credited with saving at least seventeen passengers, repeatedly plunging into the water to haul survivors to shore. Spencer collapsed from the ordeal, asking all who were present "Did I do my best?" For the rest of his life, Spencer was an invalid from his rescue efforts. When he was honored by Northwestern in 1909 with a bachelor's degree in the arts, Spencer appeared on the commencement platform in a wheelchair.[81] Spencer is memorialized by a plaque at Northwestern, and by the Christian hymn *Have I Done My Best for Jesus?* During his life-saving efforts, Spencer at times swam with a rope tied to his body. Someone stood on shore holding the rope until Spencer reached a survivor, and then the holder used the rope to pull in both Spencer and the survivor. The person who held the rope and pulled Spencer and a survivor to shore was Northwestern professor Henry Bannister—young Henry Martyn's father, and Edward Spencer's professor at the Garrett Biblical Institute.[82]

At the time of the *Lady Elgin* disaster, Kennicott was exploring the upper reaches of the Mackenzie River. He may have learned of the disaster through correspondence, and likely learned more about the event when he returned to Chicago, Evanston, and the Grove in 1863 because of the ill health and subsequent death of his father. Kennicott also knew the impact the Civil War had had on Evanston and Northwestern. His cousin, Warren Kennicott, enlisted in the Union Army and was killed at the Battle of Gettysburg. After the shelling of Fort Sumter, a meeting at the Methodist Church in Evanston held the evening of April 23, 1861, saw "the entire village" congregate, including all Northwestern faculty and students. Students came forward and "vied with each other in signing the patriot's pledge." When a call for volunteers went up, "young men we know and esteem," including some "beardless boys, with their slight forms and flushed young faces," came forward. "Cheer after cheer went up from the excited audience as each one took the pen and wrote his name as a volunteer in the Army that goes to save the Union."[83] Henry Martyn Bannister was one of only two graduates in the Northwestern Class of 1863, along with Almus Butterfield, and the youngest member of his class. He was sixteen years old when he, his father, and his family attended the 1861 war meeting at the church. At least eight students from Bannister's class left Northwestern prior to graduation, all enlisting in the Union Army. Butterfield enlisted soon after graduation, leaving young Henry the only member of his Northwestern class not to enlist in the Union Army during the war. Henry's father in 1863 delivered a funeral eulogy in Evanston memorializing Lieutenant Henry Meacham, a former Northwestern student who volunteered and then died of disease while serving in the Union Army. Had Henry attended this memorial service, he would have heard his father attribute the death of Meacham to the fact the victim was "frail."[84]

Henry's father, who visited William Dall at the Chicago Academy of Sciences on occasion to talk about his son Henry, who held the rope for Edward Spencer during the *Lady Elgin* disaster, and who appeared at church with the other Northwestern faculty for the 1861 war meeting, was in fact concerned about his son. He wondered about young Henry's stamina, courage, and other attributes often associated with notions of masculinity. Whether and how the father directly expressed these concerns to the son is unknown, but correspondence from the father hints at these concerns. In 1862 the elder Bannister wrote Alexander Winchell at the University of Michigan. He told Winchell "I have a son nearly 18 years of age, now, in the Junior class here, whom I would like much to have under your care and instruction. . . . His mind runs to science . . . he is small . . . a youth of excellent morals, but timid and shrinking, and needs to encounter men and see more than, in this quiet place, he can see."[85] After graduation, instead of enlisting, Bannister joined Kennicott at the Smithsonian. Young Henry's appointment at the Smithsonian also was guided by the pen of his father, who had written Kennicott about the matter. Kennicott agreed "there can be no doubt as to the propriety of your sending Henry on at once. . . . [L]et me have a fair trial with him for some three months."[86] The next day, Kennicott wrote directly to young Henry, advising him to bring "a pair of strong boots and suits of coarse clothes if you have them—I shall take you into the *woods* next spring."[87] In the first few days of 1865, the Bannister family began corresponding about Henry joining Kennicott and others on the Western Union Telegraph Expedition.[88] As news spread through Evanston, Northwestern University professor Oliver Marcy wrote young Henry with congratulations as well as reminding Henry of his responsibilities on the expedition. "I am rejoiced at the opportunity you have to make such a trip. . . . You are at the right age—have had no inferior education. . . . You are under patronage of the great monopoly of all science in this country and I think certainly you will be at fault if you do not succeed."[89]

Edward Spencer and the *Lady Elgin*, the war meeting at the church, his classmates, the Evanston townspeople and their unflagging support for the Union, his own father, and one of his former Northwestern professors all contributed to accelerating young Henry's trajectory into manhood. It turned out that an appointment to the scientific team came along with the patronage of more than one great monopoly. The members of the scientific team all received Union Army commissions by Western Union prior to departure from San Francisco. Kennicott had earlier considered enlisting in the Union Army but eventually hired a substitute. In August 1863 Dall spent five days guarding the Watertown Arsenal outside Boston with other volunteers.[90] Having avoided military participation on the Civil War battlefields, Kennicott, Bannister, and Dall nevertheless found themselves military officers of an unusual sort as a result of the Western Union Telegraph Expedition. The blurring of Western Union and the Union Army, with the Civil War as a catalyst and the telegraph as a bonding agent, created conditions that transformed the three naturalists into three military officers. The military naturalists, along with other USMT officers involved in the expedition, were the first American military officers on the ground in Russian Alaska for an extensive period. They set the stage for the influx of the American military into Alaska even before the 1867 Alaska Purchase.

Political versus Environmental Masculinities

Disdaining comments are sprinkled throughout the extant diaries, journals, and correspondence of Kennicott, Bannister, and Dall about their experiences with the USMT officers and Western Union men of the expedition.

It is reasonable to conclude the three naturalists had little regard for the practices of comradeship, adventure, and masculinity as conducted by their military cohort. Instead, they found comradeship in the mutual commitment to natural history, adventure in fieldwork, and masculinity in the challenges of surviving months, if not years, in little-known lands, climates, landscapes, and ecosystems. For Kennicott, Bannister, and Dall, the proving ground of bravery was not the battlefields and engagements of the Civil War, but the landscapes and ecosystems of the Arctic. They sometimes referred to themselves and each other as *carcajou*, French for wolverine, the smartest and most cunning survivalist of all the boreal animals. Their masculinity was not political but environmental.[91] All three knew that the boreal ecosystem was a different zone of conflict: not a zone of bullets, artillery, and bayonets. The boreal, subarctic, and arctic landscapes and ecosystems were zones of survival amid a never-ending array of natural dangers, any one of which could lead to injury, sickness, starvation, exhaustion, malnutrition, and ultimately death. Every single day in the field meant facing the question "did I do my best?" for oneself, for each other, for the scientific component of the expedition, for the museums, and for natural history. For Kennicott, Bannister, and Dall, their lived experiences of danger and masculinity were not based on the experiences and attitudes of Civil War officers and soldiers. They were closer to something like the disaster of the *Lady Elgin* and the ordeal of Edward Spencer: life-threatening encounters with an unforgiving environment that could ultimately lead to physical damage, infirmity, and even death. Yet one could not simply duck and hide away, for to do so would put at risk the goal of natural history. One had to be like Edward Spencer in Lake Michigan, but do so without Professor Bannister on shore, holding the other end of the rope. In the Arctic, there was no rope.

And so it was that in the waning weeks of 1865, William Dall returned to San Francisco with orders to continue running the scientific team, send thirty-two boxes of specimens to the Smithsonian, secure new supplies, and return to Alaska in 1866.[92] Henry Bannister, rescued by a native Alaskan, narrowly survived freezing to death and Kennicott stationed him at St. Michael Redoubt. Robert Kennicott, as he had predicted in earlier correspondence with Baird, took a small party up the Yukon in the heart of an arctic winter to collect specimens and scout potential telegraph routes. In 1866 Bannister kept a meteorological journal. Kennicott himself died in May. Knowing none of this beforehand, Dall returned to St. Michael Redoubt on September 25, 1866. He noted in his diary that as landfall neared he had "a feverish anxiety for which I cannot account, to see Kennicott."[93] He learned the news when a boat came out to meet them in the harbor. On September 30 Dall, Bannister, and others placed Kennicott's coffin on board a departing ship. Bannister would return while Dall stayed. The snow blew all around them as they prepared the ship for its voyage and eventual arrival in San Francisco. Entering the events of the day in his diary, Dall protested "leaving syphilitic men" behind and was then reprimanded for his protest by a USMT officer. Disgusted with both his military cohort and the Russian hosts, Dall wrote "Damn these double faced hounds. . . . Fisher drunk as a fool sleeps in my room all night. Stepanhoff has just raped the only virtuous girl in the fort."[94]

Dall chose to stay in Alaska and continue the research of the scientific expedition, while Bannister sailed on to San Francisco.[95] Dall also knew—but was never able to tell Kennicott—that a June 1866 fire at the Chicago Academy had destroyed many of the arctic and subarctic specimens previously collected by Kennicott and others. What no one stationed in the Arctic with the Western Union Telegraph Expedition knew in the last few months of 1866 was that an Atlantic cable had been successfully relaid. Continuous and reliable undersea cable transmissions between North America and Europe were flowing by September 1866, a few weeks before Kennicott's body was shipped off from Alaska on its long voyage home to the Grove. Kennicott had written Spencer Baird in 1865 that if the expedition had enjoyed efficient organization, proper supplies and provisions, and a timely arrival in Alaska, there was good reason to believe the line could be completed by October 1866. Kennicott's estimate of completion turned out to be within a month or two of the successful establishment of a working and reliable Atlantic cable.[96] The disorganization of the Western Union Telegraph Expedition meant that the Bering Strait cable was not to be. Dall first learned of the continuous success of the Atlantic cable on July 24, 1867, when he returned to St. Michael Redoubt to see the American flag flying on a mast. He learned, "The news is as follows. The United States have bought Russian Alaska for $7 million in gold. The Western Union Telegraph Co. has suspended operations in Rus. Am. on account of the complete success of the Atlantic Cable & all hands are going home. Baird is at the head of a Scientific Expedition now at Sitka."[97] Some American flag-hoisting in Russian Alaska had already preceded the 1867 purchase agreement. Expedition members at Nulato hoisted the American flag on April 23, 1866, when Kennicott returned to Nulato from an excursion. One expedition member on the excursion noted that as Kennicott and their party neared the fort at Nulato, "what should we see but our large 15 foot American flag floating in the breeze, it was really cheering."[98] Alaska was an American preoccupation for many expedition members prior to the 1867 purchase and eventual transfer.

Telegraph construction crew in Alaska, 1865–66 (Dall-Healey family photographs, William Healey Dall carte de visite album, ca. 1858–1911, "Group of unidentified men posed near stone fence, mountains in background," n.d., photo 69.92; Collection of the Massachusetts Historical Society)

The various telegraph construction crews in Siberia, Alaska, and British Columbia largely returned home as the news of the Atlantic cable reached them. Construction crews in Siberia, pondering what to do with unused equipment and wire, "opened a sort of international bazaar" and "cut the price of telegraph wire until that luxury was within the reach of the poorest Korak family." They also successfully sold shovels, pickaxes, glass insulators, and other supplies to indigenous buyers.[99]

Dall stayed on alone in Alaska until August 1868. In Washington, Henry Martyn Bannister unpacked and cataloged some of the Alaska specimens that had arrived at the Smithsonian.[100] Bannister was one of the only persons in the city who had ever been to Alaska, and he provided information about Alaska to Congress as part of the hearings about the Alaska Purchase. He also was one of the few people in Washington at that time who was reasonably fluent in Russian.[101] His meteorological journal dispelled notions that Alaska was icebound the entire year.[102] The Russian Extension was never completed, although poles were put up and line strung in several locations. Nevertheless, the Russian Extension and the support of Congress in 1864 had ensured military

Sketch of Native Alaskans, 1866 (Alaska State Library, Archives, and Museum, Daniel B. Libby Sketchbook Collection, identifier ASL-MS0107-73-2-1)

security for the Transcontinental Telegraph and other Western Union lines throughout western North America. Western Union telegraph lines across the Great Plains and elsewhere now under military protection and security reaped massive profits for Western Union and returned significant dividends for Western Union investors. Put simply, Western Union's plans to build telegraph lines and cables *beyond* the borders of the United States proved crucial to ensuring military support and protection for many Western Union telegraph lines *within* the borders of the United States. The ambiguous mix of the military and the corporate produced a valuable dividend for Western Union and for commercial telegraphy on the North American frontier.

The Americanization of Alaska

The Alaska Purchase brought the physical territory of Alaska into American possession, and the purchase also delivered data and information. The Alaska Purchase included a wealth of Russian cartographic and ethnographic information about Alaska that arrived in Washington as part of the purchase.[103] Dall returned from Alaska to the Smithsonian in December 1868 and in 1870 published *Alaska and Its Resources*. From 1871 to 1880, working with the Coast and Geodetic Survey, Dall returned to Alaska for further research.[104] The Great Chicago Fire on October 8–10, 1871, destroyed the Chicago Academy of Sciences and presumably put an end to any lingering thoughts Dall might have entertained about returning to Chicago. Kennicott was gone, and now so was Kennicott's "Young Smithsonian."

View of Sitka, Alaska (Dall-Healey family photographs, William Healey Dall carte de visite album, ca. 1858–1911, n.d., photo 69.147; Collection of the Massachusetts Historical Society)

But Alaska was now part of the United States of America, despite the lack of an undersea cable to Russia. In July 1868 Chief Signal Officer Albert Myer asked the Department of Alaska's general Henry Halleck to detail an officer under his command for Signal Corps training.[105] Halleck visited Alaska, taking along Eadweard Muybridge, who had been appointed by the army to photograph Alaskan landscapes, cities, and indigenous peoples. In the early 1870s Lucien Turner served in the Signal Service at various locations in Alaska and later became a leading naturalist and ethnographer for the Smithsonian Institution. Turner's report from the Signal Service station at St. Michael Redoubt (renamed Fort St. Michael after the 1867 purchase) was one of the most detailed reports from any of the hundreds of stations reporting in 1874. It held information on weather conditions, aurorae, astronomical phenomena, the growing seasons of plants and vegetables, bird migrations, Native Americans and Russians, trappers, animals, and fishes.[106] Turner was eventually replaced at Fort St. Michael by Edward Nelson. Nelson, who briefly attended Northwestern in 1875, became a leading Alaska expert and a preeminent American ornithologist in his generation. He wrote Henry Martyn Bannister in 1879 from Fort St. Michael that he had at last "fulfilled my promise" to write and that much had changed. Apologizing for his epistolary brevity, Nelson wrote "hurriedly" so he could send the letter "on the supply ship of the North Pole Expedition," closing with "regards to my friends in Evanston."[107]

Meteorological reports and measurements from Alaska and the Arctic expanded from the work of Henry Martyn Bannister and became vital for Signal Service weather forecasts and analysis of the Arctic, North America, and Europe. Polar expeditions were launched in the 1870s, climaxing with the 1881–84 Lady Franklin Bay Expedition led by Signal Corps officer Adolphus Washington (A. W.) Greely. Work in Alaska, the Arctic, and the subarctic regions proved important for both national and international weather maps. In an 1881 report, the Signal Service recognized the "ocean is the birthplace of storms and on the Pacific are propagated the great storms that sweep eastward." As a result, the "prime object of the American meteorologist" was "to secure the earliest intelligence" of storms moving across the North Pacific Ocean and from that data "be able to foretell the extent of their development when they cross the vast plateaus which flank the Rocky Mountains and parallel ranges."[108] The development of daily national weather maps by the Signal Service in 1871 and the inauguration of international weather maps in 1878 in part resulted from data gathered from Pacific and northern regions of western North America. Sharing weather data for daily maps began in November 1871 with Canada and with Mexico in 1877. In 1875 the publication of the *Bulletin of International Simultaneous Observations* began, which included maps in its issues starting in 1877.[109] The weather map soon became "a symbol of international scientific cooperation."[110] In the 1879 annual report of the Signal Corps, Albert Myer proclaimed that the inauguration of a daily map of the Northern Hemisphere in July 1878 was "without a precedent" in the history of the nation and demonstrated the shared pursuit of knowledge among the "civilized powers of the world north of the equator."[111]

In the 1870s A. W. Greely was among the leading planners, organizers, and inspectors of Signal Corps and Signal Service military telegraph line construction in the Dakotas, Texas, and the Southwest. The 1881–84 Lady Franklin Bay Expedition that Greely led attempted to lay the groundwork to establish an American colony in Greenland and also allow American polar explorers to claim a "furthest north" for the United States, part of a growing interest and competition among nations to claim the furthest north on the way to eventually reaching the North Pole. Beset by financial irregularities and also frozen in for three years by sea ice, the Lady Franklin Bay Expedition was tragic and sensational, captured national and global press cover-

Telegraph line in Alaska, late nineteenth century (Collection of the University of Alaska–Fairbanks, Edward R. McFarland Photographs, identifier UAF-1974–130–50)

age, raised the specter of cannibalism, and made Greely a national hero. He would go on to command the Signal Corps in the 1898 Spanish-American War, organize the army relief efforts in the aftermath of the 1906 San Francisco earthquake, and be a cofounder of both the Explorers Club and the National Geographic Society. As Chief Signal Corps officer in 1900, Greely secured funding for and launched construction of the Washington-Alaska Military Cable and Telegraph System, later known as the Alaska Communications System.[112] A few months before his death in 1935, Greely received the Congressional Medal of Honor for "splendid service." Greely and Charles Lindbergh are the only individuals to receive the Medal of Honor for actions other than bravery and valor on the battlefield.

It was Adolphus Washington Greely who finally stood as the American apotheosis of Arctic masculinity. During his era, Greely became the avatar of both the masculinity of an officer and the masculinity of a naturalist, cementing the political with the environmental. In 1891 Greely, now a general and Signal Corps commander, delivered a farewell address to the

Alaska, 1921 (Anchorage Museum, H. G. Kaiser, Alaska Railroad Collection, image number B1979.002.AEC.G1729)

many meteorologists and weather observers of the Signal Service as weather observation duties transferred out of the Signal Corps into the newly established National Weather Bureau in the Department of Agriculture. He noted that his duties as a Signal Corps officer in the 1870s and 1880s had brought him "into unusually close relations with its enlisted force in almost every section of the continent, from the valley of the Rio Grande to the plains of Dakota and Montana, and in the dreary wastes of the polar regions." What he learned from these many encounters was that experiences with challenging landscapes, ecosystems, and climates "necessarily develop the character of men." Greely stated he learned the lesson—"over twenty years' service, covering the prime of his manhood"—that Signal Service duties were "fraught often with danger and hardship" and the ultimate beneficiaries of such duties were "the nation and the Army."[113]

Reminisces and Repatriations

Dall and Bannister for the most part went their own ways. Dall eventually returned to the Smithsonian and became the foremost Smithsonian expert on Alaska and a leading naturalist of his generation. Bannister eventually returned to Evanston, served as a curator at the Northwestern natural history museum for a few years and later went on to a career in medicine, specializing in insanity and mental health issues. The Signal Service ran a training office on the Northwestern campus for one year in 1875 but abandoned the effort when students balked at attending three times a day, throughout the day and night, to monitor weather conditions.[114] Lifesaving boats and crews, however, remained active at Northwestern. The federal government built a lifesaving station on the Evanston campus in 1876. On Thanksgiving Day 1889, the Northwestern student life-saving team rescued the entire crew of the *Calumet* in "one of the fiercest blizzards known in that region in years." The team received the Gold Medal for distinguished conduct and bravery by the secretary of the treasury, the highest honor awarded by the U.S. Live-Saving Service. Appeals over the years to award a medal to Edward Spencer went unheeded. The Coast Guard took over the Northwestern station in 1916 and in 1931 moved the operation to nearby Wilmette Harbor. Eventually, what little remained of the Kennicott collections at Northwestern and the Chicago Academy of Sciences made their way to the Field Museum.

Late in life, Henry Martyn Bannister lived out his final years in Evanston. His pastor, Horace Greeley Smith, called the elderly Bannister "the most unforgettable character I have met" and, in a speech honoring Bannister, began by identifying Bannister not by name but as "The Man in the Wheelchair." Smith explained that Bannister, in part as a result of his Alaska ordeals, now had reached a point in his life where "a strange and incurable malady has laid hold of him—the kind that cripples and weakens, but does not destroy or kill." Despite his travails, Smith knew that Bannister "never allowed his crippled condition to embitter him."[115] Smith added that Bannister read Richard Dana's *Two Years before the Mast* "almost every year." As a young man in Alaska, Bannister had often sketched sailing ships and other vessels in his logs and diaries.

Reminiscing about his Alaska experiences in a 1909 interview, Bannister saw his Alaska past in part through his medical training and research. He recalled that of "all the parties that were sent up to Bering Straits and vicinity, there were but few deaths although there were at least two suicides and as many or more cases of insanity brought on from the exposures." Thinking about the need to properly prepare and equip for extreme environments, Bannister pointed to the toll the lack of proper preparation could take. "Of our own party of fourteen, one became insane before we came home and at least one more after reaching civilization and both cases I think can largely be attributed to the conditions undergone." Bannister believed he would have "escaped any serious consequences at all if I had been properly equipped."[116] He discreetly hinted at the dysfunctional management of Bulkley, the USMT officers, and Western Union. He reiterated the warnings Kennicott gave long before and again upon arrival in San Francisco in 1865. Kennicott, along with Dall, repeatedly advised Bulkley and others while provisioning in San Francisco that the equipment and gear was improper for the Arctic and unsuited for the pursuit of telegraph construction and for the pursuit of scientific research. Western Union and the USMT officers failed to properly equip men for work in the Arctic. They also failed to provide those men with adequate nutrition while working and surviving in Arctic conditions.

Bannister returned from Alaska still a young man. Over the years and with personal experience, careful reflection, and the knowledge gained from medical training, Bannister eventually recognized something else that had happened in Alaska. Late in his own life and looking back to his Alaska past, he saw in the experiences of some of the men on the expedition the signs of what now is often called post-traumatic stress disorder, exacerbated by the problems of dysfunctional military expeditions. Perhaps the Western Union Telegraph Expedition was one of the earliest examples of a long-term American overseas military presence. If so, then perhaps the expedition made manifest a disease associated with dysfunctional military occupations. Perhaps the Western Union Telegraph Expedition also was a military occupation of Russian Alaska. One way or another, at least some expedition members were continuously on the ground in Alaska from the initial arrivals of August and September 1865 until October 1867, when Alaska became American. Simone Müller accurately noted that the nineteenth century was a new era for the engineer as a symbol of civilization, taking on a role that "appeared to be even more distinguished than that of the military, the traders, or the missionaries, who had in former centuries played the central role in first encounters."[117] In this sense, the engineers—in this case, the leaders of telegraph exploration and construction—joined the ensemble of actors previously involved in occupying foreign lands, and in the case of Russian Alaska, did so hand-in-hand with military officers and natural historians.

Viewed as a military occupation, or preoccupation, the roughly two-year period was, at that time, the longest continuous American military presence on foreign soil to that date. The on-the-ground presence in Russian Alaska was continuous from the arrival of the expedition until the transfer of power. The expedition turned out to have some of the experiences now more strongly associated with long-term American military presence on foreign soil: the problems of supply and provisions, the challenges of local climates and environments, ambiguous concepts of the central mission, and perhaps for some the experiences and lingering damage of post-traumatic stress disorder. Was Alaska occupied before it was purchased? Not in terms, conditions, practices, and policies clearly related to contemporary experiences of military overseas occupation. Long-term American military presence and occupation on foreign soil is, however, a historical trajectory that emerged from somewhere, likely in bits and pieces, not full-blown but with a few hints and clues as the first instances began. Perhaps the Civil War, the USMT, the Smithsonian Institution, the Western Union Telegraph Expedition, the Chicago Academy of Sciences, Northwestern University, and the Alaskan experiences of three young natural historians—Robert Kennicott, William Dall, and Henry Martyn Bannister—provide some hints.

After a very long journey home, the body of Robert Kennicott was interred at the Grove in a grave next to that of his father. The obituary in the *Western Monthly* reported that the funeral services for Kennicott took place "during one of the coldest days of January" in 1867. The mourners were accompanied by "birds (particularly the black-cap titmouse) in great numbers on the trees." As the coffin was lowered, "a flock of quails approached and gave forth a call" and then in chorus burbled "a plaintive note" as if the flock "desired to sing a requiem over his grave."[118] Bannister died in 1920 and Dall died in 1927. First organized as an American district, Alaska formally became an American territory in 1912 and achieved statehood in 1959. The Grove, Robert Kennicott's boyhood home, was designated a National Historic Landmark in 1976. In 2001, the curators of the Grove exhumed the remains of Robert Kennicott. A Smithsonian scientific team conducted a forensic examination at the Grove in hopes of determining a cause of death.[119] The examination concluded that Kennicott had died of heart failure, ending rumors and suspicion about other possible explanations. In 2005 the Smithsonian Institution repatriated to Nulato the human remains of a native Alaskan collected by William Dall, and in 2009, repatriated to nearby Koyukuk the human remains of a native Alaskan collected by Edward Nelson.[120] After the forensic examination of Robert Kennicott, his remains were not reinterred at the Grove. At the invitation of the Smithsonian forensic examination team and with the agreement and support of Kennicott family descendants, the remains of Robert Kennicott were repatriated to his other home: the Smithsonian Institution. His skeleton is now part of human anatomy collections at the Smithsonian's National Museum of Natural History. May all the repatriated rest in peace.

5

Hot Winds on a Sun-Baked Desert

The Southwest

The rainbow has but a bad character: she ever commands the rains to cease.

—Zuni weather proverb[1]

Where was the water? Answering that question for the landscapes and ecosystems of western North America was and is a central issue for human success, culture, sustenance, and even survival. To know the local water was to master local knowledge. Adequate rainfall meant crops and pasturage. Lakes, rivers, and watersheds meant fish, wildlife, transport, irrigation, edible plants, and possibly timber. Water attracted birds. Too much water at once could be dangerous, especially when winter snow melts or sudden squalls flooded rivers and washes and sometimes triggered avalanches or mudslides. But scarce water—no rivers, no lakes, no fulsome watersheds, scant snow melt or rain, not even a heavy morning dew—scarce water as the everyday condition of a landscape or ecosystem is a challenge to survival. Knowing the water in the landscapes and ecosystems of habitually dry climates is local knowledge of the most significant order. Such knowledge enhances the likelihood of daily survival. Nowhere on the North American continent was (and is) this endless experience of an arid climate and extreme temperatures more typical of everyday life than the landscapes and ecosystems of the American Southwest. It was no wonder that when the Zunis gazed upon the rainbow, they saw a prism of character flaws. The Zunis and other indigenous peoples of the American Southwest were unparalleled masters of North American water knowledge, whether of rain, snow melt, irrigation, morning dew, or the locations of rivers, streams, creeks, and—most important—the waterholes. To know the waterholes meant survival.

Frank Hamilton Cushing, the Smithsonian anthropologist who collected the Zuni weather proverb opening this chapter, represented the rise of a new collecting protocol for the Smithsonian. Unlike Robert Kennicott and the earlier Smithsonian emphasis on the collection of natural history specimens from the North American frontier, Cushing and other Smithsonian anthropologists active in the Southwest during the 1870s and 1880s collected stories, songs, daily habits, objects, religious and ceremonial artifacts, oral histories, and other attributes of Southwestern indigenous culture. Cushing eventually became a member of the Zuni and was, in a distinct way, somewhat of a Zuni master.[2] Albert Myer, founder of the Signal Corps, was certainly no Zuni master, but he was familiar with the American Southwest, having begun his military career in 1854 at Fort Duncan (Eagle Pass), Texas. He later moved on to Fort Davis, where his duties as an army doctor included weather observations (as was common for army medical personnel at that time).[3] While traveling from Corpus Christi to Eagle Pass, Myer wrote about his experiences. Myer found "a prairie (devoid of) natural beauty" with nothing but grass "now all withered." There were "bushes whose entire structure seems to be of thorns." The search for water was ongoing and often fruitless when found. "Even water fails you and unless you would drink that fluid from the same mud hole which mules and men have defiled you must carry your supply in your keg." Myer later added, "it never rains here. . . . I write in the midst of dust." No matter the location "the whole earth is parched." Signs of rain or snow were always, it seemed, dashed to disappointment. "There will be no snow . . . there has been no rain . . . (some days) the morning dawns cloudy. Now, you think, comes the rain, but by noon the sun is shining and at night you see it setting in the same clear sky that has been over you for weeks before." Myer summarized his letter by stating that "nothing grows in western Texas" and consequently there were "no articles of produce. Our lumber comes five hundred miles, our corn a thousand."[4]

Military Telegraphy in the Southwest

Twenty-five years after his postings at Fort Duncan and Fort Davis, Albert Myer was chief signal officer of the army. When General George Crook arrived in Arizona Territory in 1871 to assume command of the Department of Arizona, he immediately recommended a military telegraph line be built from California to Arizona. Territorial governor Richard McCormick had made a similar recommendation in 1865 and again in 1868.[5] Myer's annual report of 1879, as did his previous annual reports, discussed the military telegraph and the western frontier. By 1879 the military telegraph lines planned and constructed for "Arizona, New Mexico, and the Texas frontier were now

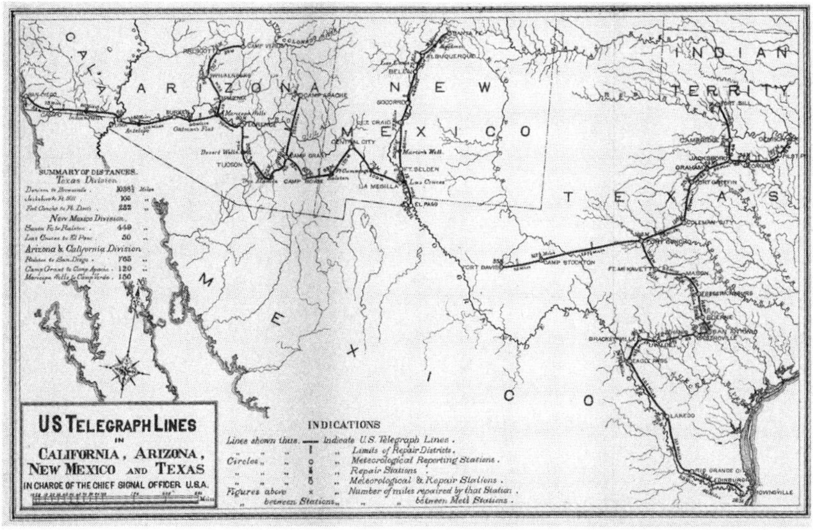

An 1878 map of military telegraph lines (*1878 Annual Report of the Chief Signal Officer*, p. 188)

completed." He added that the country "where such lines pass has but few points where water could be had." This was important for surveillance, containment, and potential conflict with indigenous peoples. During such hostilities "the line of Indian retreat, the points they must pass to reach water, are approximately known" meaning that troops could "occupy the passes and points" during conflicts. "There can be no constructions more important for holding a frontier or protecting the first steps of advancing civilization than the telegraph lines." Troop movement directed by telegraph "attained results which could not, without the rapid action made possible by the lines, have been hoped for." In Santa Fe and elsewhere, "telegraphic communication, both east and west, has been more secure and constant than at any previous period." He added that these military telegraph stations in the American Southwest also furnished regular Signal Service meteorological reports.[6]

For Santa Fe and other stations, this was a marked improvement in military telegraphy over that of the early 1870s. Santa Fe began filing Signal Service weather observations in November 1871, corresponding with Washington by postal routes and also by telegraph "with as much regularity as the condition of the telegraph line would permit." Santa Fe observations were usually sent by postal systems, which incurred problems because postal communication with Santa Fe was "unreliable" with "frequent delays."[7] In 1873, the Santa

Fe station found telegraphic communication with the Signal Service network "very unreliable" and both weather monitoring and the coordination of troops were hindered as a result.[8] By the end of the 1870s, many of these problems of telegraphic communication had been overcome. The challenges of the forbidding high and dry landscapes and ecosystems of the American Southwest were successfully met by the 1880s. The signal flow became more reliable for all manner of data, including the conveyance of state secrets in the form of battle orders for American military troops carrying out Native American surveillance, combat, and containment. The secure command of American electronic information networks as a military asset in the 1880s blossomed in the deserts and arid landscapes, bolstering territorial white settlers in the Southwest while suppressing its indigenous peoples. The water became known by the military and settlers in some ways similar to those of the original inhabitants, and in other ways shaped by a different local knowledge focused on control and suppression of Native Americans.

The initial 1859 experiments by Albert Myer in military signaling with flags, torches, and lanterns were carried out in the Southwest during the Navajo campaign. As the telegraph arrived in the Southwest in the 1870s and 1880s, the area once again became a testing or proving ground for new tactics and uses of military communication, particularly uses of the military telegraph in engagements, conflicts, and surveillance of indigenous peoples. The telegraph became an important component in border control with Mexico, assisting in the tactics of hot pursuit across the border and in the extradition of suspected criminals. Preemptive tactics, the profiling of persons of interest, rapid response, and the synoptic surveillance of Native Americans as they moved about beyond the borders of newly established Indian reservations all involved the ongoing use of the telegraph. The military telegraph lines in the Southwest were increasingly deployed as mobile assets. Deployment involved the routine rerouting of extant lines to deal with new developments, the installation of lines for military control near geographically strategic locations such as trails and waterholes, and the regular use of temporary tie-ins to telegraph routes by military personnel in the field, who carried a telegraph key, sounder, and wire.

In these ways, the military telegraph in the Southwest in the last two to three decades of the nineteenth century presaged the new uses and testing of military technologies that continued in the Southwest throughout the twentieth century. On the heels of the telegraph came new military communication and transportation technologies such as the heliograph, the field telephone, bicycles and motorcycles, flashlights, balloons, and motor vehicles and airplanes. Eventually, in the latter half of the twentieth century, rockets and missiles arrived in the Southwest for testing and deployment. Telegraph networks giving early warning of adversarial actions in the Southwest were the ancestors of global electronic information technologies, networks, weapons, and systems for national security, such as the "distant early warning" (DEW) line in northern Alaska, Canada, and Greenland or the North American Aerospace Defense (NORAD) Command. In July 1945 a new secret weapon developed in New Mexico at Los Alamos and tested at Alamogordo lit up the sky even more spectacularly than had the Great Prairie Fire of January 1865. The military telegraph at the end of the nineteenth century was a precursor and progenitor of a long and continuing period of new military tactics and technologies tested, developed, and on occasion deployed in the American Southwest.

Prior to the arrival of the military telegraph in Arizona and New Mexico Territories, communication via overland mail, stage routes, and couriers usually occurred at a slow pace, even a crawl. Military orders issued to the territories by mail and courier from the Department of California could take three months or more to reach the various posts and forts in Arizona Territory. In comparison with California, "St. Louis was considered to be as near in time, if not closer; papers and letters from that city reached Tucson in eighteen to twenty-one days."[9] The asymmetric communication had a deleterious impact on a whole range of military activities, from conflict and containment of indigenous peoples to provisioning forts and stockades to promotions and courts-martial. Further, the arid climate and intense heat exacerbated supply problems. Military-issue saddles quickly succumbed to damage from the desert conditions, as did wagons. Mules, horses, and livestock struggled with the local climates and environments. News accounts and letters from elsewhere arrived late, if at all. Most military personnel detested living in these forbidding conditions of harsh environments and extraordinary isolation.

By 1870 military reorganization by the army removed Arizona Territory from the Department of California and designated the territory as the Department of Arizona, allowing more autonomy for military commanders within the territory. The reorganization also alleviated in a bureaucratically typical way the long duration of communicating department-level military orders by mail, stage, and courier. When General George Crook arrived in 1871 to take command of the Department of Arizona, in his initial reports he stressed the need for rapid communication by telegraph. Crook reported that the Department of Arizona was isolated from California and elsewhere.

It faced challenges of military cohesion. Rapid communication between troop deployments was crucial in effectively containing and controlling indigenous populations. A telegraph connection to California as well as a telegraph network within Arizona Territory would be necessary and central factors in effectively addressing these challenges. Other army commanders agreed with Crook's assessment, particularly as it pertained to conflict and containment with indigenous peoples. Generals John Scofield and William Sherman seconded the telegraph proposal and added that a state of war existed in Arizona Territory. By 1872 plans were underway to establish telegraph connectivity between San Diego and Arizona Territory and also within the territory.[10] By the middle of November 1873, the military telegraph line was up and running between San Diego, Yuma, and Prescott, followed by further construction of branch lines in Arizona Territory. Telegraph lines reached Tucson in December. Tucson was interconnected to the east with lines reaching Ralston in New Mexico Territory by May 1877.[11] These perspectives on the necessity and value of the telegraph as a military asset routinely spurred requests for additional lines and operators from military commanders throughout western North America in the 1870s and 1880s.[12] Whether the military telegraph operated in the Southwest, the Great Plains, or the Dakotas and whether deployed against Apaches, Sioux, or Nez Perce, the telegraph increasingly shared information about the ongoing wars waged against Native Americans. This included General Philip Sheridan in Chicago, General William Sherman and the War Department in Washington, and many other military officers and government officials. As Elliott West observed, the growing military telegraph capacities spreading across the Southwest and across the full range of western North America meant "General Sherman oversaw it all from wherever he happened to be."[13]

Arizona settlers shared a perspective with the Minnesota settlers who had petitioned Congress in 1850 to extend a telegraph line to Minnesota Territory: they "considered it the business of the Federal Government to build the telegraph." A military telegraph in their locale would "atone for the past neglect" of Arizona Territory settlers and ensure settlers would no longer be left "to the mercy of the Indian and Mexican cut-throats and robbers" preventing peace, progress, and civilization. The telegraph also enhanced border control and protection "should war with Mexico or any other foreign country come on us." A telegraph network would greatly aid in the surveillance and, when needed, coordinated military and law enforcement action against indigenous peoples when their activities appeared to be threatening, suspicious, or hostile.[14] Arizona settlers also agreed with others in the mili-

Vigilance court in session (Denver Public Library Digital Collections, call number Z-3804)

tary, in Congress, and at commercial telegraph companies that there was not yet sufficient business activity in Arizona Territory to justify and generate revenue to profitably support private investment, construction, and operation of telegraph lines. Some argued the lack of sufficient commerce to support commercial telegraph routes was not sufficient justification to deny the benefits and security of the telegraph to Arizona settlers and other settlers of the Southwest, because the arrival of the telegraph also meant the end of settler isolation.

Albert Myer was sympathetic to this perspective when he argued in favor of building military telegraph lines in West Texas in 1874, despite the challenges of the landscapes and ecosystems and the lack of immediate promise of commercial activity. The arrival of the telegraph in the Southwest would ensure that "little settlements will grow up" and military posts and personnel "will no longer be isolated." The arrival of the telegraph also would accelerate the arrival of new settlers. "The inducement to enterprise, to carry settlement forward will cease to be counteracted by the thought that such settlement amounts to self-banishment."[15] Secretary of War William Belknap wrote Congress in 1874 that the establishment of military telegraph lines in West Texas would provide further protection for frontier settlements "against Indian and Mexican depredations," and the War Department supported the expansion of military telegraph lines throughout the Southwest. Belknap added that military telegraph construction in the Southwest had the support of army generals, and in an optimistic note, mentioned there was a "probability of such lines being self-sustaining" by returning revenue from telegrams sent by white settlers and local commerce.[16] Military telegraph line construction in Texas took place with the support of troops at several forts. Troops stationed at Fort Davis built telegraph lines on several occasions between 1874 and 1880. These activities and new telegraph lines meant the army could increasingly "coordinate its efforts from its telegraph stations at Fort Concho, Grierson's Spring, Fort Stockton, Fort Davis, and Fort Bliss" on Native American containment and border security. The construction efforts came with a cost, for military telegraph line construction "strained frontier resources."[17] If the increasing number of commercial telegrams on military telegraph lines in Arizona Territory served as an indicator, the "little settlements" did grow after the arrival of the telegraph with the increase in commercial telegraph activity throughout the latter 1870s and into the 1880s.[18]

The end of settler and military isolation in the desert Southwest became a centrally agreed upon justification for bringing Arizona and New Mexico Territories into national telegraph networks, including via the military telegraph. This was discursively grounded in issues of national security, protecting settlers, and border control and surveillance of, containment of, and conflict with Native Americans. Indeed, it later turned out that as soon as a telegraph line reached an Indian reservation in Arizona and New Mexico Territories, "the telegraph assumed a new function—instant news of Indian defections."[19] This "instant news" was the product of synoptic surveillance of indigenous peoples, a development that brought an adapted vision of the theory and practice of synoptic weather observation by the military into the synoptic surveillance of human beings by the military. Frederick Godfroy, the Indian agent at the Apache Mescalero Reservation, had "a telegraph line to Fort Stanton, usually in condition for use."[20] Some suggested the telegraph would increase amity between Arizona settlers and military troops. By creating conditions that could likely induce additional national newspaper coverage of Arizona and facilitating more in-depth news reports from within the territory into national circulation, it might even ameliorate "much of the misrepresentation regarding civil and Indian affairs in Arizona which has puzzled the eastern press and confused Congress, the Departments, and the public for some years past." Finally, stories and tales from the recent past had reached settlers in Arizona Territory from the Great Plains and elsewhere across western North America, reinforcing the faith and belief of white settlers in the "psychological effect" of the telegraph on indigenous peoples. Such stories mythologized for new white settlers in the Southwest how the telegraph lines were often left undisturbed by Native Americans because of their "suspicious regard" of the telegraph. The prevalent white attitude of Native American technological primitivism spread to additional regions of the North American frontier.[21]

Weather, Landscapes, and Ecosystems in the Southwest

Another factor motivating the increase of military telegraph lines in the Southwest was the growth of the Signal Service and the collection of weather data. The Signal Service station in Galveston, Texas, stood ready in 1872 to issue precautionary warnings and storm predictions for the Gulf of Mexico, as did other coastal stations on the Atlantic and Pacific seaboards. Despite preparations for the display of cautionary storm signals in Galveston, no storms had been locally reported as of 1872 "owing to a want of a sufficient number of reports from points in the northern and western parts of Texas to afford a reasonable basis of observation."[22] For Galveston in particular, advance warning of storms from the north was crucial in coordinating cotton shipments. This meant knowing which points "in the interior are visited or effected by 'northers' *before* their arrival in Galveston. The reports from Fort Gibson, Shreveport, Denver, and other places will show this." Advance storm warnings were crucial because "*after* a 'norther' of any force strikes here ships cannot leave the wharves while it lasts."[23] As the first weather reports arrived, albeit irregularly, from additional Signal Service observation stations in the Southwest by 1874 and 1875, the promise was seen for the future. "While but scanty reports have as yet been had . . . the utility of those received is manifest."[24] In Galveston, Indianola, and elsewhere on the Gulf Coast, more reports from the interior were wanted for accuracy in storm predictions. Too many opportunities for timely storm warnings about the Gulf of Mexico were lost because "telegraphic interruption" prevented the collection of weather reports from elsewhere.[25] By 1876 forty interior military telegraph stations in Texas and in Arizona and New Mexico Territories collected and telegraphed daily weather observations to St. Louis, San Diego, or Santa Fe for further transmission to Galveston and the Gulf Coast and on to Washington.[26] Visitors to the Galveston Signal Service weather observation office found the station weather observers increasingly helpful on impending "norther" storms, particularly in the winter season, when "reports from the military telegraph line enable them to judge quite accurately when to expect these storms, and they time and shape the direction of their trips accordingly."[27]

The expansion of the military telegraph lines and addition of military telegraph stations in the Southwest served double duty (or could be seen as a dual-use technology) by augmenting weather data collection and coordinating military actions, all the while enhancing the surveillance of nature and the surveillance of indigenous peoples. Albert Myer wrote in his 1879 annual report about the dual-value surveillance made possible by military telegraph lines and Signal Service weather observation stations. "The events of the past year have illustrated what might be the uses of such lines in Indian wars. They make safer the settlement of the country. They make possible the establishment of stations valuable for meteorological reports."[28] The telegraph was increasingly becoming a central component of multiple American synoptic surveillance systems. Timely collection, assessment, and dissemination of relevant data transmitted by telegraph increasingly made surveillance a synoptic process regarding the weather of Western North America as well as of the indigenous peoples of Western North America. These telegraphic dynamics of multiple synoptic surveillances had local, regional, national, and continental manifestations.

Work began on military telegraph line construction in the summer of 1873 for a telegraph line with a western terminus in San Diego and a territorial terminus in Yuma, with further lines from Yuma interconnecting several locations in Arizona Territory. An early problem arose with planning the route and the unreliability of extant maps. Reliable maps of Arizona Territory, the southeastern section of California, and the border with Mexico proved to be in scarce supply. The few maps available were often twenty years old or more and contained topographical uncertainties and old abandoned roads. Particularly vexing was precisely determining the border with Mexico, because for security reasons the telegraph line from San Diego to Yuma needed to be wholly contained within the United States.[29] The border with Canada had proven on several occasions to be telegraphically porous for both commercial and military lines, yet interconnectivity between Canadian and American telegraph lines and networks nevertheless flowed amicably.

The issues with American telegraphy on the Mexican and Canadian borders were wholly different in the 1870s and 1880s. Unlike the Canadian border, where the telegraph was usually a gateway to shared United States commerce and weather data between the two nations to the north, American telegraph lines on the Mexican border needed to be able to extend to and exchange

Nogales, Arizona, with telegraph line on the American side of the border with Mexico, 1898–99 (National Archives and Records Administration, American West Photographs, number 163 Nogales Arizona, 48-RST-3A-18)

messages with Mexican telegraph lines. More important, the telegraph needed to serve the United States as a crucial component of a secure wall against border incursions from the nation to the south. Such a secure wall came to pass in the 1880s. The military telegraph station in Campo, California, reported in 1881 their location near the border with Mexico was valuable in exchanging telegraph messages with Mexican authorities and meant the telegraph "affords quick means of applying for assistance when the place is threatened by border ruffians." Citizens and settlers believed the "speedy means of communication gives a feeling of security that would not prevail were the telegraphic facilities removed." The only regret of the settlers was insufficient commercial telegraph business in the area "to recompense the government for the great benefit the community derives from the line."[30] As the 1880s drew to a close, it was "apparent to every soldier that Arizona and New Mexico could not be effectively protected against Indian outbreaks without the greater part of these military lines, and the value of the Rio Grande line" near the border with Mexico avoided serious complications that would otherwise continually arise to erode border security.[31]

Mobile Electronic Communication with the Telegraph

In the 1870s and 1880s, the army and the Signal Corps further developed extensive military telegraph networks in West Texas, Southern California, and New Mexico and Arizona Territories. Officers such as Benjamin Grierson paid particular attention and carefully surveyed and reconnoitered landscapes. Grierson often accompanied troops and patrols on survey missions and pursued knowledge of maps and trails, as well as potential hideouts for criminals, trails used by Native Americans, and waterholes. The establishment of telegraph lines allowed dispersed troops to stay connected and enabled field reports by smaller parties using temporary tie-ins on patrol away from forts and stockades, which proved important in many skirmishes and encounters.[32] Commercial telegraph companies and railroads also began to build and expand telegraph routes in Texas and the Southwest.[33] Telegraph routes and connections in these territories were always built with multiple considerations: connectivity, communities, businesses, settlers and ranches, the availability of timber, military efficiency, border control, Native American surveillance and containment, and access to or surveillance of trails, streams, creeks, and waterholes. For example, the military telegraph station in Aleman, New Mexico Territory, reported its location was "of great benefit" as "two principal Indian trails across the Jornada del Muerto are situated, respectively, 3 miles below and 7 miles above this station. Several times during the year the crossing of hostile Indians" was reported by telegraph, which led to the occasional use of the station as a headquarters for the district military commander.[34] Fort Cummings in New Mexico Territory was near the "most important trails used by the hostile Apaches" and as trail sightings of indigenous peoples were made, Apache movements were "promptly communicated to district headquarters, to troops in the field, and to citizens and settlers, enabling them to take prompt action whereby many lives and much property has been saved."[35] Newspapers in Arizona Territory and elsewhere expanded their coverage with telegraph reports and published more national news stories because of their new access to telegraph news agencies such as the Associated Press.[36] Telegraph news agencies and individual commercial and military telegraph stations—combined with temporary military telegraph field stations during campaigns against Native Americans and against unwelcome individuals moving back and forth across the Mexican border—allowed army officers to deploy a version of mobile telegraphy as a tactic of containment and interdiction. There had been occasional examples of military telegraph mobility during the Civil War and during the conflicts on the Great Plains in

1865–66, but the mobility of the telegraph and the tactic of temporary tie-ins from field patrols and zones of conflict became routine military actions in the American Southwest and surpassed earlier deployments.

When troops were on patrol or in the field, the telegraph moved with the officers in command and with troops and patrols. They consistently maintained connectivity to other field patrol units and to Fort Davis, Fort Concho, and other forts, stockades, and garrisons in western Texas and the Southwest. Of course, these locations were further connected to the Signal Corps's nationwide telegraph network, and it was not unusual for telegrams from field patrol reports to eventually make their way to the War Department in Washington. The actions and tactics of mobile telegraphy in the American Southwest were local and contingent on the moment, and the telegraph reports and data transmissions were received and analyzed locally in the Southwest as well as nationally in Washington. Military telegraph routes in Texas, Arizona, and New Mexico changed from time to time and poles were dug up and relocated and lines were restrung.[37] Old forts were closed, new forts were built, new white settlements took hold, and Native American reservations were established and patrolled. The military telegraph in the Southwest was not a static but a flexible infrastructure.[38] On occasion, Signal Service weather observation stations also were moved. The Signal Service telegraph and weather observation office in Wickenburg moved several times between 1875 and 1887.[39] (Although some telegraph lines and routes of western North America moved, some routes to forts, stockades, and larger settlements such as Tucson, Phoenix, or Santa Fe remained fixed.)

Temporary tie-ins from anywhere in the field to main military telegraph lines could be added to the flexible infrastructure and such tie-ins became routine actions. Virtually every patrol, even a single scout or soldier, carried a tie-in wire, sounder, and telegraph key. The military telegraph network in the Southwest became the initial proving ground for the earliest military applications of mobile electronic communication networks in surveillance and conflict. The Signal Corps recognized in its 1877 annual report the importance of coordinating telegraph lines with known waterholes and other locations of water, particularly in the American Southwest. The coordination of military telegraph lines with sources of water was a key to success in combating and containing Native Americans. "A telegraph line well worked forms one of the most efficient of barriers against the raids of Indian war parties. . . . There can be no constructions more important for holding a frontier or protecting the first steps of civilization than the telegraphic lines."[40] Telegraph lines built near waterholes and in areas believed to hold the potential for conflict were, when possible, built with iron poles, in part to prevent vandalism by burning the poles. Benjamin Grierson used iron poles whenever available when building telegraph lines near waterholes in 1879.[41]

The importance of monitoring waterholes and reporting activity to various forts and garrisons by military telegraph was clear during the 1880 military campaign against the Apache warrior Victorio. Victorio and the Apaches in his command had, on and off for two years, stymied both Mexican and American troops with speed, stealth, and multiple border crossings. At first, Mexico and the United States had difficulty agreeing on a reciprocal plan for their troops' hot pursuit of Native Americans, criminals, and suspect individuals back and forth across the Mexican-American border.[42] Telegraphic connectivity between American and Mexican troops was intermittent, with connectivity interrupted for several weeks at one point by severe storms. Despite missed opportunities and the lack of a clear reciprocal agreement for hot pursuit, the monitoring of waterholes in both Mexico and the American Southwest eventually led to a conclusive fatal attack on Victorio by Mexican troops.[43]

Troops employed the military telegraph throughout the 1880s as an asset in Native American conflict in the Southwest. Victorio and other Apache warriors stymied military efforts against them by routinely cutting telegraph lines. Some reports mentioned the sabotage of telegraph lines by Native Americans, who would cut a line and then replace the cut line with a strip of leather or something else that looked like a telegraph line. Such acts were much more difficult to rectify, with no readily identifiable cut and downed telegraph line. Repairs thus meant inspecting long stretches of wire to find the fake line. Previously Apaches had cut telegraph wire and "miles of it had been thrown into deep canyons." Apaches "sent the boys up the trees to cut the wire, tie the ends to limbs with buckskin thongs, and conceal the breaks." This was done "wherever they found the wire attached to a tree."[44] In many cases the army was able to quickly work around such acts and reestablish telegraph connectivity among forts, stockades, and troops in the field. Sometimes all it took was stringing a new stretch of wire. By the 1880s telegraph wire had long ceased to be a scarce commodity and was manufactured and distributed in abundance, as were insulators and other telegraph components. Railroads eased the problems of delivering telegraph wire.

Some examples from the 1880 Signal Corps Annual Report are illustrative. On the "9th of April, Private Falvey, while waiting orders here, accompanied Captain Kramer's command, Company E, Sixth Cavalry, as a scout. . . . He took with him a key and relay, and each night, as the trail lay near the line, cut in and placed Captain Kramer in direct telegraphic communication with headquarters . . . and for two weeks transmitted many important messages over the wires." Fort Grant, Arizona Territory, found that "During the late Indian outbreaks commanding officers have frequently established their headquarters in telegraph offices, and through the medium of the operators have transmitted orders and directed the movement of troops as well as if they had been present at any point communicated with. By means of the telegraph it has been possible to quickly concentrate the forces, to aid and protect the settlements, and to engage or pursue the hostiles." Camp Thomas, Arizona Territory, advised that "The great convenience and utility of the telegraph line has been clearly demonstrated during the recent raid made into this section of the country by hostile Indians." The mobile military telegraph could also, presumably, promote peace, according to officers at Wickenburg, Arizona Territory: "during the military operations against the Chemjuevis Indians in May last, when this line became the principal medium of communication between the department commander at Prescott and his subordinates and agents in the field and at this point. Full instructions and lengthy detailed reports were all exchanged at short intervals, enabling all parties to act with promptness and intelligence, and thus avert what threatened to be a bloody Indian war."[45]

The Need for Standardized Time

Although an American electronic information network in the Southwest was becoming increasingly flexible, adaptable, secure, and reliable, one of the issues that remained vexing in the desert, indeed, across the entire expanse of the continent, was time: to be specific, the lack of standardized coordinated time zones. During the 1870s, the Signal Corps and the army were among the many national constituencies—scientists, railroads, farmers, merchants—increasingly concerned about the problems inherent in the practice of local time determinations. Signal Service observers were advised that "it frequently happens that the telegraph time differs slightly from the local standard time" and to check their own clocks against the clock of the "telegraph manager."[46] The Signal Service station in Virginia City, Montana, observed the July 29, 1878, solar eclipse "assisted by sixteen citizens" who closed their businesses to participate, who also donated some funds for "the necessary apparatus" to study the eclipse. Even with this local assistance, the reporting station apologized for some of its inaccuracies in measurement. "On account of getting Washington time but once, our numerous records were not as accurate

Bowen, Colorado, in the late nineteenth century (Collection of Colorado History, Stephen H. Hart Library and Research Center, call number CHS.X4302)

as they would have been had the telegraph facilities been as good as at other points where observations were taken." Despite these problems, "some of our work was very successful, especially one of our corona sketches, which, according to the opinion of experts, could not be excelled."[47] A map of the path of the eclipse, along with twenty-eight sketches of the corona, was published in the 1881 Signal Corps report.[48]

The Signal Corps shared its eclipse observations with the Naval Observatory. The Virginia City Signal Service station confirmed their time records "can not be of much value, as we got Washington time but once, and that not satisfactory" and sketches were supplied because there were no photographers in the vicinity. Edward Berthoud, chief engineer of the Utah and Northern Railroad, reported to the Naval Observatory that he used his chronometer and observed the eclipse about forty miles west of Virginia City. He admitted his time for the beginning of the eclipse "may be 2 or 3 seconds out, as some Indians at that time left precipitately and somewhat drew my attention for a moment."[49] The eclipse attracted many observers to the eclipse path, including Thomas Edison. Edison tested his tasimeter to measure the corona heat during the eclipse with an eclipse team stationed at Rawlins, Wyoming.[50] Lucien

Turner had no tasimeter, chronometer, or telegraph station with him when he was posted at the Signal Service station at Fort St. Michaels, Alaska, in 1874 (the same location where Henry Martyn Bannister took meteorological readings in 1866). Rather than attempt to log his observations on Washington time, Turner "adopted local noon" and regulated his clock at Fort St. Michaels by "a sun-dial of Russian construction."[51] When Ferdinand Westdahl was in Alaska on the Western Union Telegraph Expedition, he found the planet Jupiter was difficult to observe "on account of the planet's great southerly declination." He also had difficulties with local latitude and longitude observations. The North Star was "too high over the horizon" to be effectively used with a sextant. There were endless "compass needle variations" and to top it all off, Westdahl admitted he could not "regulate my watch."[52]

Precise measurement often seemed a challenge in western North America, whether it be Ferdinand Westdahl stymied in the Arctic, Edward Berthoud startled by the movement of Indians, or the dangers to troops measuring the water levels of the Missouri River. Robert Kennicott, William Dall, and Frank Hamilton Cushing seemed not to have such problems, yet they were less involved in measurement and observation, and more centrally involved in collecting specimens, objects, and stories. Nevertheless, the problems of precise observation and measurement at many locations across the North American frontier during the latter decades of the nineteenth century became evident on a number of occasions. They were particularly manifest in the determination of the precise difference between local time at any given location in western North America and Washington local time. Railroad time often was used, although railroad time across a locale or region could vary with the particular railroad line. Synoptic surveillance of a multitude of locales was compromised by the imprecision in—and indeed impossibility of—synchronizing operations because of the lack of coordinated time.

The problems inherent in the lack of coordinated time were well known to the leading meteorologists of the Signal Service. Cleveland Abbe and other meteorologists joined the growing chorus of scientific voices calling for coordinated or standard time. In particular, as Ian Bartky noted, "in the 1870s scientific pursuits requiring simultaneous operations from scattered points became important" and standard time was crucial for effectively carrying out simultaneous scientific observations and measurements across a large geographical expanse. Standard time coordinated across widely scattered points of scientific observation and measurement was a key for gathering data about important astronomical events such as eclipses. The interest of Abbe and other meteorologists in studying the aurora borealis of April 7, 1874, spurred Abbe to begin calling for adoption of standard time.[53] In 1875 Abbe chaired a study committee on standard time for the American Meteorological Society. His 1879 committee report advocated using the current system of longitudinal meridians. The report advised that any steps to establish standard time needed the support of corporations such as railroads and also needed political support. Abbe encouraged Congress to take steps to promote international agreements to establish global standard time.[54] Bringing time zones and temporal standardization across the North American continent into the daily routine of Signal Service weather observations and data collection would increase the likelihood of more precise weather predictions and storm warnings by Signal Service meteorologists.

Standardized time needed above and beyond all else precise determination of longitude across the entire span of a given time zone. Longitude had been a challenging computation for centuries. The development of the marine chronometer in the 1760s enabled relatively precise longitudinal determination while traveling on the open seas. The chronometer remained a crucial instrument into the twentieth century for determining the longitude of one particular location (or plotting a series of longitudes along a voyage). But even an array of carefully calibrated chronometers placed at several points across a span within an area would not be sufficient to precisely determine longitude for each and every point within that area. This fact was well known to scientists and also to telegraphers, who realized that for scientific research and for telegraphic communication involving synoptic observations and reports over various points and distances, only those observers and telegraph stations coincidentally located near a useable clock for synchronization could effectively participate. Even here, to choose the synchronized time of one railroad line, for example, would not necessarily precisely coincide with the synchronized time of a competing railroad. Decisions needed to be made in order to choose a single time measurement standard for coordinated measurement and synoptic precision. Time needed one and only one railroad, or much better yet, one and only one clock. This meant one and only one astronomical observatory as the standard by which all the clocks for all the synoptic measurements across all the areas and distances would thenceforth be calibrated. "Clearly, the issue of multiple times had to be resolved."[55]

In 1872 Western Union president William Orton told Congress the telegraph companies knew the "impossibility of making twelve watches, however carefully adjusted, hung on the wall, and at the same temperature, keep time together." They also knew the difficulty of having telegraph operators at different stations "all so quick of ear that the same click of the instrument (not always as clear and precise on one instrument as another) conveys to each at the same time the same idea."[56] The human auditory system lay at the center of the simultaneous key-and-relay auditory-tactile telegraph process. Humans, no matter how well trained and adept at their key-and-relay skills, were like a collection of twelve watches hung on a wall. They were individually reliable and accurate, but not uniformly precise in operation. Nevertheless, the telegraph and the chronometer working together could achieve sufficient longitudinal precision. Determining longitude for a specific locale came about through repeated measurement by two observers, followed by the application of the "personal equation" used in accounting for the differences in observational practices and reaction times of the two different observers measuring at the same location. Local measurements of this kind were calibrated by telegraph with an astronomical observatory.

As early as June 1844, U.S. Navy officer Charles Wilkes conducted a one-time experiment using the Baltimore-Washington Morse telegraph line to determine Washington longitude. "Wilkes reckoned that the Capitol in Washington was 1 minute 34.868 seconds of time west of Battle Monument Square in Baltimore."[57] Into the 1850s and beyond, the telegraph became a central instrument for astronomers and others working in geodesy. Richard Stachurski noted that astronomers and scientists "followed the wire" to plot longitude across North America and also followed the cables as they were laid beneath the oceans and seas. The Atlantic cable in conjunction with the growing telegraph networks of North America brought the Greenwich Observatory into North American geodesy. By 1880 the accumulation of several years of surveys, observations, and computations across North America by the Coast and Geodetic Survey had "made 127 telegraph measurements connecting a network of stations that extended from Greenwich along the East and Gulf Coasts to Galveston, Texas, and across the midsection of the country to San Francisco and Seattle via Omaha and Salt Lake City."[58]

Despite the growing significance of the Greenwich Observatory for global longitude and geodesy, until 1912 the Washington Meridian passing through the Naval Observatory was used as the American meridian for state and territorial boundaries. This "uneven differential" from the Greenwich Meridian accounts for the fact that the western boundaries of Nebraska "are not expressions of whole degree numbers today."[59] Thanks to nationwide telegraph lines and transatlantic cables, it was the Greenwich Meridian that shaped the standardization and coordination of time for the North American frontier and across the planet. Observations and measurements were taken in western North America at several locations, and each location brought its own particular conditions. The longitudinal measurements done by telegraph at longitudinal calibration points set up in Cheyenne and Colorado Springs during

1872 and 1873, for example, had to take into account factors affecting access to telegraph lines. In 1872 Western Union furnished a branch line to the Cheyenne calibration station and provided a company telegrapher to work at the station except in cases where there was sufficient commercial telegraph business at the Cheyenne telegraph office throughout the day and evening that "did not permit him to leave, which sometimes happened." In such cases an off-duty telegrapher served the survey as a substitute. When the telegraph calibration signal reached Salt Lake City, the fact that "weather-signals are sent by telegraph each evening about 6 p.m." and that Salt Lake City also relayed these and other area weather reports eastward had to be accounted for.[60]

Determining longitude and the ongoing practice of geodesy would prove important in the establishment of standardized time and coordinated time zones in the United States in 1883. The global impetus for longitude by telegraph and cable enabled the global measures taken to introduce time zones in the 1880s. President Chester Arthur, in the crowning foreign policy achievement of his presidency, convened the 1884 International Meridian Conference in Washington, creating a common ground and procedures for nations to coordinate longitude and time zones.[61] Coordinated time zones came to the nation and to the telegraph. The precise determination of longitude for the North American frontier also reinforced the ability of the federal government to precisely determine the boundaries of Indian reservations. No longer

Green River, Wyoming, in the late nineteenth century (Denver Public Library Digital Collections, call number Z-5795)

largely constrained or in part forced to determine reservation boundaries by local landmarks or changes in the terrain, the telegraph brought longitude and latitude to western North America. This produced a sufficient exactness that allowed the federal government for the first time to delineate reservation boundaries and borders with little, if any, regard for local rivers, canyons, trails, forests, meadows, lakes, or mountains. Surveyed boundaries precisely denoting latitude and longitude, boundaries then marked by allotment stakes, became the means of delineating the perimeters of reservations. The telegraph delivered the science of geodesy to the frontier, shattering the sensibilities of local landscapes and ecosystems for indigenous peoples. Their daily lives would henceforth be increasingly lived and experienced within the perimeters of geodetic shapes construed upon federal maps shaping and guiding governance over the landscapes and ecosystems of western North America.

Telegraph as Social Media: Chatting on the Line

In addition to new approaches and practices in geodesy, determining longitude, and standardizing time, new attitudes and practices toward the routine use of the telegraph by military officials and others in the Southwest and elsewhere emerged in the 1870s and 1880s. Various military officers and post commanders in Arizona and New Mexico Territories as well as West Texas communicated with each other constantly by telegraph on military matters. They also fostered social friendships and professional relationships in what Torsten Kathke called an atmosphere that "was constant, loquacious, and free-wheeling, with some commanders quite drawn to near-instant messaging."[62] The local and territorial telegraph practices of military officers in the Southwest took on attributes of continuous messaging, messaging while traveling as well as when at one's post, short texts with regular updates, and a community sharing in daily electronic dialogue.

Such chatter and bantering among military and government telegraphers was not confined to the Southwest. It was another bone of contention for Western Union with the Signal Service and its weather observers. In October 1872 Western Union president William Orton complained to Albert Myer that special arrangements and new circuits established by Western Union for the Signal Service were being "put to other uses than the correction of errors in reports" and related weather data matters. Orton protested that these circuits "were being taken for purposes of display, during the most crowded hours of the day" by Signal Service observers and operators "to give exhibitions to invited guests, and to promote the convenience or gratify the whims of its attachés." Orton included copies of Signal Service messages sent between Signal

Service operators over these special Western Union circuits asking who might be visiting a particular city and how the telegraph lines of Western Union competitors were performing. Additional examples of such dalliances collected by Western Union were congratulatory messages, character references, gossipy items about other telegraph operators or offices such as "they were a kind of self-governing people in Green Bay," queries about smallpox, a request for the name of "the best occultist in Baltimore," and local train schedules.[63] Increases in the total mileage of telegraph lines and in the total hours those lines were in use spurred increases in the incidental uses of the telegraph in the 1870s and 1880s by military personnel and others. Such uses were somewhat akin to contemporary social media practices such as texting and messaging during other activities. Others frowned upon these habits then as now, particularly when done at the workplace on company hours using company equipment.

Another example of an innovation in the Southwest was the new heliograph—mirror signaling. Heliograph systems, when properly manufactured, calibrated, and operated, allowed mirror signaling over great distances in elevated areas. In the Arizona conflict against Geronimo in 1886, General Nelson Miles made "urgent telegraphic requests" to Signal Corps commander William Babcock Hazen for "men expert in the use of the telegraph."[64] Hazen sent telegraphers and also experts in the heliograph. The indigenous peoples of the American Southwest also used mirror signaling. James Kaywalka, an Apache who as a young boy knew Victorio, recalled that Apache sentinels "were provided with bits of mirrors to flash warnings" and that General Nelson Miles was mistaken in his belief that the Apaches "knew nothing of the heliograph." Kaywalka said his "people did not know the Morse Code, but they had learned that mirror flashes warned soldiers and ranchers of our movements" and therefore Apaches also "made use of the device."[65] John Pershing, who was involved in many military campaigns against Native Americans early in his career, was among the officers involved in heliograph tests.[66] Pershing subsequently served in Cuba during the Spanish-American War and in the postwar occupation of the Philippines. He was an observer in the 1905 Russo-Japanese War and an observer in the Balkans in 1908 and fought during the Moro Rebellion in 1913. Pershing led the 1916 Mexican Punitive Expedition and was commander of the American Expeditionary Forces to Europe in World War I.

A different example of Mexican-American border control and the telegraph emerged in 1892. That year, the Signal Corps constructed and operated a temporary or "flying telegraph" line along the Rio Grande about one hundred miles west of El Paso. The temporary telegraph line was built by the Signal Corps across "a very difficult country . . . devoid of water . . . scanty

Las Vegas, New Mexico, in the twentieth century (Denver Public Library Digital Collections, call number L-492)

vegetation . . . numerous outcroppings of rock . . . soil and conditions most unfavorable . . . the heat was oppressive . . . conditions of dust and drought most trying." The line was in use for only four evenings between May 14 and 17 and was used by the International Boundary Commission "for the purpose of determining, by telegraph, the longitude of the monument marking a point upon the international boundary of the United States and Mexico . . . with results eminently satisfactory to the Commissioners."[67] International cooperation in weather forecasting, undergirded by synoptic observations collected by telegraph, was one example of the telegraph and scientific measurement in western North America. Standardized time emerged as another example, later joined by the third example of boundary determination, whether it be the boundaries of Native American reservations or the border with Mexico.

Transformations in Federal Science Funding

In the last decades of the nineteenth century, the continuous interactions between scientists, the telegraph, and the military in the American Southwest and western North America began to diminish. Congress was increasingly dubious of the management and quality of the Signal Corps, in no small part because of the 1881 embezzlement of Signal Corps funds by Henry Howgate, exacerbated by Howgate's 1882 escape from prison. (Howgate was not recaptured until 1894.)[68] The Signal Service itself was one of several military-scientific operations of the federal government that in the 1880s was increasingly subject to a larger question about the relationship of various government scientific organizations to the army and the navy. Voices grew in favor of the transition to civilian agencies for federal government activities in weather forecasting, surveying and geodesy, geology, hydrography, and other areas. With the diminishing construction of new military telegraph lines, the weather operations of the Signal Service by the 1880s had become the largest single component of the Signal Corps. A military unit in the army, commanded by a general, was in practice becoming a unit less centered on military affairs and more centered on meteorological affairs. From many perspectives, more and more questions emerged about whether the Signal Service and its weather forecasting belonged in the army.

The weather question was one of a number of similar issues investigated in Congress from 1884 to 1886 by the Allison Commission. Named for its chair, Senator William Boyd Allison of Iowa, the commission was "an *ad hoc* joint committee of the United States Congress" established to "examine the organization of the scientific bureaux of the federal government." The formal name of the Allison Commission described at length its investigative charge: The Joint Commission to Consider the Present Organizations of the Signal Service, Geological Survey, Coast and Geodetic Survey, and the Hydrographic Office of the Navy Department. These military agencies accounted for about $3 million in annual appropriations and represented the major federal budget expenditures for scientific research.[69] Some saw the Signal Service ideally as a civil, rather than military, scientific agency. The Signal Service was military in orientation primarily because of federal budgetary decisions rather than decisions that weather data and analysis was primarily a military (rather than scientific) concern. Cleveland Abbe, trying to find wiggle room and aware of occasional suspicions about government-funded research endeavors, called the Signal Service "a piece of applied science" and gamely trotted out the comment that "the British Army had done more meteorological work than the U. S. Army."[70]

The Allison Commission split on the central question of whether weather forecasting should remain with the army. Many had suggested the Department of Agriculture as a more suitable home for American meteorology. Chief Signal Officer William Babcock Hazen successfully argued to retain meteorology in the Signal Corps until his death in 1887. His successor,

A. W. Greely, took a completely opposite tack and helped create the conditions, including advising on legislation, for the Signal Service to leave the Signal Corps and join the Department of Agriculture as the Weather Bureau, later the National Weather Service and now the National Oceanic and Atmospheric Administration. President Benjamin Harrison signed legislation on October 1, 1890, approving the transfer of government meteorology from the War Department to the Department of Agriculture. The Signal Corps was out of the weather business.

Public and congressional opinions about military and scientific endeavors had changed. The universities had displaced West Point and Annapolis as the leading institutions of scientific and engineering training and research. Many no longer saw the need or value in having the military front and center for government research projects and science funding. The leading experts were now civilians with established professional careers in a wide array of scientific and engineering specialties. Congress and the Allison Commission wondered why military budgets, usually funneled through House and Senate committees on appropriations, were the major federal funders of science. Congress saw opportunities to move science and engineering budgets away from the military branches into other branches and agencies of the federal government. That would mean other congressional committees would have greater influence and oversight over those civilianized federal research endeavors. There were those here and there who bristled at the idea of a military man becoming a man of science. Perhaps it was a question of mission, perhaps of masculinity, perhaps the early sentiments of a coming progressive era, perhaps skepticism about progress. When William Babcock Hazen died in January 1887, the *Cincinnati Enquirer* rued the fact that some had taken umbrage at Hazen "because of his literary nature" and admitted that regrettably some people took "a kind of wild bison dislike of a man who could write a letter or a book."[71] By the end of the 1880s, a consensus had emerged. The military was not always the proper place for federal funding of scientists, meteorologists, and researchers. According to a few, military science risked battle-tested Civil War veterans softening over the years into literate eggheads. The Department of Agriculture would now be an active government user of American telegraph networks. The Signal Service became the National Weather Service, and with the transfer of Signal Service personnel to the Department of Agriculture in 1890–91, the number of officers and troops in the Signal Corps declined precipitously.

As the new Chief Signal Officer, A. W. Greely saw the looming crisis for the Signal Corps. He admonished the lack of personnel and budget support in the aftermath of the Signal Service transfer. "The telegraph, the heliograph, the electric flash light, and the telephone have become potent factors in civilized warfare, and the continued neglect of a military corps which is charged with their study, practice, and operation, cannot commend itself to any prudent or thoughtful citizen."[72] Just before the transfer of weather responsibilities to the Department of Agriculture, the Signal Corps had begun experiments with avian-borne messages. "Experiments have been made, from Key West as a basis, during the year with homing pigeons. The liberality of prominent bird-fanciers" helped to acquire pigeons "at a minimum of expense" to the army. Birds were trained to fly at intervals up to one hundred miles over open water.[73] The risks of flying over open water was not a challenge in the Southwest but was a concern in coastal areas.

For the Signal Corps, weather data via the telegraph was gone. Twenty years of one central and daily experience of the military and the telegraph in western North America had been carried away by the changing zephyrs wafting through Congress and the general public. But for the military, the telegraph, and the North American frontier, the preceding thirty years, starting with the Transcontinental Telegraph, had presented another set of central and daily experiences: surveillance, containment, and long-term conflict with indigenous peoples. As the nineteenth century closed, the Signal Corps would rise again to the occasion of war and conflict. The coming war of 1898 would, like a war fought from 1846 to 1848, be a war against a foreign power and a war once again yielding new territories for the nation. Unlike the war with Mexico, however, when President James Polk was completely isolated from the battlefield and his generals in the nadir of telegraphic connectivity, fifty-odd years later President William McKinley would enjoy ongoing telegraphic connectivity with the military at the White House. McKinley exercised the secure command of multiple electronic communication networks at war, now at their zenith. The White House and the nation had arrived at the moment when the United States could deploy the many lessons learned over the decades from the telegraph and the North American frontier to the world at large. These nineteenth-century telegraph lessons about surveillance, secrecy, security, ecosystems, weather data, the high ground, secure command, measurement and precision, standardized time, and warfare were about to be wired into nature over all Planet Earth. American electronic communication networks were going global.

Conclusion

For the first time in American history the president, in the fullest sense, was enabled from his executive office to exercise his prerogative as Commander-in-Chief.

—Harold Bolce, "The Telephone at the White House—The President's Private Branch Exchange"[1]

The White House Telegraph Room (also called the Telegraph and Cipher Bureau) of the William McKinley presidency was the room where "the President has the whole United States under his thumb."[2] In 1877, at the request of President Rutherford Hayes, the Signal Corps assigned Benjamin Montgomery to the White House as a telegraph officer. For the next twenty-eight years, Montgomery oversaw the expansion of the White House Telegraph Room into what one reporter writing in 1902 for *Telegraph Age* called "perhaps the most complete bureau of confidential communication in the world."[3] During the Spanish-American War, Montgomery created the facilities for President McKinley to maintain cable and telegraph connectivity with the fleet in Manila and direct battlefield communication via cable and telegraph with troops in Cuba.[4] McKinley credited Montgomery and the White House Telegraph Room with helping to halt an ongoing battle by flashing telegraph messages to the army in Cuba immediately upon the signing in Washington of a protocol suspending hostilities.[5] The Signal Corps rapidly updated McKinley on successful efforts during the conflict to contain and prevent international undersea cable traffic in and out of the West Indies and Greater Antilles.[6] The Associated Press assured the White House it would practice self-censorship on any wartime reporting for both military and public relations issues, agreeing not to disseminate or publish anything "likely to give aid to the enemy or to embarrass the government."[7]

Montgomery experimented with graphophone transcriptions of both telegraph messages (the sound of the key) and telephone conversations as White House security measures.[8] Montgomery was the only military officer present at the formal signing of the peace protocol ending the war with Spain (held at the White House) and appeared in official photographs of the protocol signing.[9] He installed many telegraph and telephone lines (and switchboards) to enhance the White House capacity for communication to Congress and the War Department in Washington.[10] He improved the abilities of the White House to use the telegraph and telephone as both election campaign networks and election return monitoring networks. He proved indispensable to the Executive Office and was an early version of what today might be called an intelligence analyst, electronic communication specialist, or a senior staff member of the National Security Council. "Repeatedly he has worked through the night at the White House key, with a President at one elbow and a Secretary of State at the other, manipulating the nerves of the world. For twenty-five years he has been, in effect, the President's eyes and ears."[11] McKinley exercised a secure command over an electronic communication system that increasingly proved its global viability. A Utica newspaper told its readers that the global reach of the White House electronic communication system "passes through jungle and marsh up to the very firing line in the Philippines, and it stretches through the allied armies to the walls of Pekin. It radiates through Europe to the great American embassies, and it touches the most obscure consular agency on the African coast." Readers were assured that "from the White House the President issues orders to the army and the navy in the most distant parts of the world almost as quickly as if he were personally at hand." In conclusion, readers learned that the secure command of electronic communication networks for the president meant "the whole world is his checker-board."[12]

George Squier, who would command the Signal Corps in World War I, told the officers at the 1900 Naval College course on the Spanish-American War that the war was "largely a story of coal and cables." The outcome of the war proved cables could exercise a dominating influence over the control of the seas in commercial, military, and diplomatic pursuits. Recognizing that Great Britain effectively controlled "four-fifths of all the cables in the world," he saw the British global cable system as "woven like a spider's web to include all her principal colonies, fortified ports, and coaling stations." Estimating that an American global undersea cable system could be built for twenty-five million dollars, he urged the immediate building of such a system. Closing with remarks on the need for secure American global electronic communication as a newly established world power, Squier argued that American government cables should have, as much as possible, an ambiguous undersea geography. Military officers needed to "keep the exact route of

The White House Telegraph Room during the Spanish-American War (Library of Congress Prints and Photographs Online Catalog, reproduction number LC-USZ62–90805)

important cables a secret, and prevent the publication of maps for general distribution" to the public or other nations. Squier acknowledged that despite secrecy for cable routes on ocean floors, "the shore ends, however, are certain to be known." Because of the somewhat open appearance of a cable landing site, "a cable landing, for the future, should partake the character of a fort" that provided "ample and adequate means for preventing an enemy from destroying the cable within the marine league, or indeed, until it has reached deep sea, where its accurate location is not known."[13] Squier went on to become the first passenger to be flown by Orville Wright in an airplane, develop more than sixty patents in his career, and serve as Chief Signal Officer during World War I.[14] Transcontinental undersea cables finally began to span the Pacific Ocean in the early 1900s.

McKinley was proactive in enhancing the abilities of the White House Telegraph Room. He also was a progenitor of the White House Map Room.[15] The "great map" of the McKinley White House showed longitude and latitude, political borders, empires, which nations controlled undersea cable routes, shipping lanes, major global railroads, and other global data interlaid with American commercial and military telegraph lines and undersea cables.

Montgomery was charged with running and updating this map. The maps developed by Montgomery and others essentially collapsed previous distinctions from earlier maps that delineated differences between national, continental, and global telegraph and cable networks. In essence, the Spanish-American War coincided with the introduction of global security maps for the White House. These maps projected all transportation and communication networks and factors primarily as global phenomena, rather than reinforcing the earlier distinction between the national, the continental, and the global.

Chief Signal Officer A. W. Greely reminded his officers and soldiers of this successful melding of the national, the continental, and the global in his message to the Signal Corps on September 13, 1898: "In the Cuban campaign you arranged, maintained, and operated a system of cable and land lines—partly commercial, partly war cables, partly flying-telegraph lines, and partly telephone lines—that enabled messages to pass in twenty minutes from the Executive Mansion in Washington to the headquarters of the Army." He added "throughout the war, the military telegraph lines around the great Indian reservations and along the Mexican frontier were regularly maintained and faithfully operated as in peace."[16] As Signal Corps commander during the Spanish-American War, Greely developed and carried out policies that reshaped wartime procedures for cutting cables between adversaries. These policies included cutting multiple cable landings in enemy territory and cutting cables between an enemy state and a neutral state. He also advocated and carried out "harsh censorship" of cables exchanged between belligerent parties.[17]

Montgomery received the first telegram sent to the White House from Buffalo about the 1901 assassination of McKinley. He coordinated telegraph traffic between the White House and Buffalo until McKinley passed away a few weeks after the shooting. Montgomery stayed on after McKinley's death and continued to work at the White House during the first years of the Theodore Roosevelt administration, serving as electronic eyes and ears. Montgomery was the White House staff person tasked with taking private and confidential telephone calls for President Roosevelt. Upon receiving a telephone call, Roosevelt took the call in Montgomery's private office. Montgomery also coordinated the telephone needs of the president in travel situations to other cities. Roosevelt had at his disposal telephonic connectivity with his car such that "whenever stops are made, even at small stations, the wires are quickly connected and the President enabled to talk with the White House generally getting connection with the long distance line, through the local exchange."[18] The president and the secure command of electronic communication networks became mobile.

The President Takes Command from the Signal Corps

But changes were in the air for Benjamin Montgomery and for the electronic communication networks of the president of the United States of America. In 1902 Montgomery was the only person on the Executive Office staff who refused to sign a memo agreeing not to solicit government jobs (or promotions) for himself or anyone else, arguing that as a Signal Corps officer, he was exempt from such White House staff instructions.[19] In January 1905 President Roosevelt relieved Montgomery as chief of the Telegraph and Cipher Bureau in the White House and reassigned Montgomery to a Signal Corps company command in California. This shifted command authority of the White House Telegraph Room away from the Signal Corps and firmly under the authority of the Executive Office. A local newspaper profiled Montgomery after his arrival in California and told readers "Silence is his greatest attribute." The *Washington Star* reported that Montgomery found the order to leave the White House "a complete surprise" and that the order came after a conversation between Roosevelt and Secretary of War William Howard Taft.[20] Perhaps it was something personal about Montgomery, perhaps not. It was more likely that the removal of Montgomery had little if anything to do with his personality and everything to do with his position and his status as a military officer. The removal of Montgomery as chief of the Telegraph and Cipher Bureau of the White House put an end to forty-four years of monitoring and observation by the army, the Signal Corps, and the secretary of war over the telegraph, cable, telephone, and other electronic communication activities of the president of the United States.

Since the inauguration of Abraham Lincoln in 1861, all the presidents, one way or another, conducted electronic communication under the supervision of military officers. Theodore Roosevelt put an end to that. In so doing, he changed the experience of presidential electronic communication activities such that from then on, those activities would no longer be monitored and enabled by the military, but instead secured within the office of the president. The January 1905 timing of this transition was possibly related to the results of the 1904 presidential election, which returned Roosevelt to the White House as an elected president rather than a former vice president who ascended to the office upon the death of his predecessor, McKinley. In any event, a president strongly associated with transformative visions and actions on matters of nature and the presidency proved to be a president who also carried out transformative visions and actions on matters of national security and the secure command of electronic communication for the president and the White House. Theodore Roosevelt, the great presidential champion of nature, conservation, and the environment, also transformed the office of the president to ensure full and complete presidential autonomy and confidentiality in its electronic communication activities.

By the dawn of the twentieth century, the telephone, telegraph, and undersea cable had become routine components of presidential power, including crisis management.[21] The 1902 White House renovations and construction of the West Wing saw the White House Telegraph Room move out of the old mansion and into the new facilities. The secure command of electronic communication had arrived at a location that would eventually come to be known as the White House Situation Room. The origins of the White House Situation Room are commonly attributed to the Kennedy administration, which may have been the first presidential administration to use that specific name for the facility. The secure command of American electronic communication networks for the president of the United States, however, and the establishment of a physical infrastructure within the Executive Mansion for the exercise of that secure command, preceded the Kennedy administration by an accumulation of activities by fifteen presidents spanning eighty-five years back to Rutherford Hayes and the Railroad Strikes of 1877.[22]

That the secure command was now at the disposal of the president proved evident to Waldon Fawcett, writing in 1903 about the president, the White House, and the recent conflict.[23] Fawcett told readers that the White House Telegraph and Cipher Bureau provided President McKinley with unprecedented surveillance and communication capabilities. McKinley had "a great map and secret codes" to see "every movement of the Army and Navy." Throughout the conflict he kept "in touch with men and events throughout the world." Signal Corps telegraph lines constructed in Cuba allowed President McKinley to "communicate from Washington in less than twenty minutes with the officers on the firing line at Santiago." The reach of the White House extended beyond North America and the Caribbean into the Philippines conflict and the geopolitics of the Pacific Rim. On many occasions "the President, the Secretary of War and the Secretary of the Navy sat for hours at the elbows of the telegraphers, directing in person the military operations thousands of miles distant." Projecting ahead to future conflicts and the global reach of the secure command now on the horizon for White House telecommunication capabilities, Fawcett jingoed about the possibilities of new media. He compared the recent situation of the president and the telegraph with an imagined future of a president wielding the telephone in global conflict. "Had the slender metallic strand between the President and the powder-begrimed fighters in the trenches been a telephone instead of a telegraph line, he might almost have heard the thunder of the guns and listened to the cheers of the American soldiers."

The challenges and conquest of the high ground, the need for reliable signal flow, the protection of the state secret, and the secure command of electronic information networks transcended from the domain of continental security to the domain of global security. The twenty-first-century world of a White House Situation Room, Air Force One, and the contemporary electronic communication fortress around the president, the White House, the Capitol, and the nation is a world filled with an extraordinarily vast range of electronic communication networks. It is a world that reaches back to the telegraph and western North America during the nineteenth century. The North American frontier was where the nation first wired into nature.

With the American entry into World War I, American Telephone and Telegraph (AT&T) went with the American Expeditionary Force to the battlefields of France. Soon after the conflict, A. Lincoln Lavine wrote in *Circuits of Victory* that it was a "foregone conclusion that when America entered the war, she would do most of her fighting by telephone."[24] AT&T managers and the Signal Corps had forged plans by late 1916 to establish the Signal Reserve Corps of AT&T personnel in anticipation of entry into the conflict. This Signal Reserve Corps stayed put, so to speak, during the early months of conflict, remaining in place at their jobs while simultaneously serving as Signal Corps personnel.[25] Eventually many AT&T engineers and technical personnel became part of the Signal Corps on the ground in France, instrumental in building and maintaining wartime communication systems. On the battlefields of France, the Choctaw Telephone Squadron of World War I spoke over army field telephones in the Choctaw language, effectively creating conditions of real-time encryption of battlefield communication. Recruiting the Navajo code talkers and some other Native American code talkers for World War II meant a return of the army and the Signal Corps to the people, landscapes, and ecosystems of the American Southwest. That was where Albert Myer first conceived of and tested his theories of signal communication that led to the genesis of the Signal Corps. The Choctaw and Navajo peoples were among many indigenous peoples from many Native American nations who served as code talkers for the American military in twentieth-century conflicts.[26]

Wired into Nature: A Legacy

The Native American code talkers of twentieth-century conflicts served with distinction and valor, and code talking represents a deep and lasting commitment by Native Americans to military signaling and the secure command of American electronic communication networks. The long-term trajectory of Native Americans and American electronic communication networks remains bleak, however, to say the least. Native American communities and reservations are to this day among the most underserved American populations in terms of telephony, internet connectivity, and broadband services. A 2016 congressional research service report on Native Americans and broadband services found "tribal areas and communities continue to lag behind other areas and segments of American society with respect to broadband and telecommunications services." Reasons for these conditions include the challenges of "rugged terrain" and the lack of "economic incentive" for service providers. As a result, "41% of Americans living on tribal lands lacked access to broadband at speeds of 25 Mbps download/3 Mbps upload. This compares unfavorably to 10% of all Americans lacking access to broadband at those speeds."[27]

Field biology and natural history at the Smithsonian Institution and elsewhere to this day rely on electronic communication networks for a wide array of research projects.[28] The railroads largely replaced the need for telegraph construction crews in western North America to utilize draft animals, and the railroads largely replaced the need for those crews to depend on local supplies of food. Railroad bridges and, later, highway bridges and dams made the challenges of river crossings for telegraph construction in western North America essentially disappear. The construction and maintenance of global electronic communication networks in the present day rarely if ever involves the need to use draft animals. The ability to send electronic communication and signal flow via the electromagnetic spectrum—wireless telegraphy, radiotelephony, and satellite services—often have made the need for wires themselves to essentially disappear, especially in challenging landscapes and ecosystems. With wireless, a long stretch of telegraph poles collapsed into a small number of antennae.

As of December 2016, no undersea fiber-optic cable exists beneath the Bering Strait linking Alaska and Russia. The two northernmost landings on the planet for undersea fiber-optic cable are currently Barrow, Alaska, and Longyearbyn on Svalbard Island (Norway).[29] Separated by about three thousand miles of the Arctic Ocean (about the same distance as the 1866

Atlantic cable), perhaps climate change and global warming will, some day in the future, lead to a polar fiber-optic cable beneath the Arctic Ocean linking Barrow and Longyearbyn. Such an undersea fiber-optic cable also would perhaps, in a way similar to the telegraph and the Mexican border, draw a line of a sort between the United States and Europe on one side and Russia on the other side. A direct fiber-optic cable on the Arctic Ocean seabed linking these locales would traverse the North Pole.

The NOAA and European weather agencies are huge consumers of data from global weather-monitoring stations and networks. Climate scientists also are huge users of electronic communication networks linking thousands and thousands of monitors and data points all around the world, producing a near-infinite array of measurements. Measurements and observations of landscapes, ecosystems, and nature involving contemporary electronic communication networks are far more numerous and far more complex than the measurements and observations undertaken when the telegraph first wired into nature. But this is to be expected. Measurements and methods continually change, and the techniques and technologies of measurement and observation are continually updated. Measurements, methods, techniques, equipment, collection, dissemination, and analysis of weather data gathered by telegraph across western North America in the 1870s and 1880s changed continually. New weather observation stations were established. Signal flow and circuit routing on telegraph networks was recalibrated. New categories of data were developed with methods using new scientific instruments leading to new categories of measurement. An increasing range of specialized forecasts were issued.

Beyond weather forecasts, new and changing methods and measurements related to the telegraph and the North American frontier in the nineteenth century emerged to shape and influence growing activities in surveillance, geodesy, and astronomical observations. Throwing off the yoke of local time and successfully establishing standardized and coordinated time across the continent led to yet another sea change of new measurements, methods, calculations, data collection, and analysis. The findings of the Allison Commission represented another platform spurring changes in measurements and methods. Commission findings indicated a rising agreement in the federal government and the nation as a whole that federally funded scientific pursuits were better governed and practiced with civilian, rather than military, federal agencies. The civilian scientists emergent in federal science research in the 1870s, 1880s, and 1890s brought many new ideas, measurements, and methods into federal science research, including new ideas for weather forecasting.

In 1890 the changes wrought by new measurements and methods significantly influenced another ongoing government research project involving the North American frontier, indeed involving the entire nation: the census. The 1890 census pursued many more questions than the 1880 census. Pursuing more questions involved many new methods and measurements. The 1890 census was the first census to utilize punch cards, tabulation machines, and other automated data management and analysis technologies.[30] Unlike previous censuses, the 1890 census made a significant effort to enumerate Native Americans and indigenous peoples in population records of individual states and territories. In 1860, 1870, and 1880, census researchers were instructed to include in their enumerations only Native Americans who were taxed, leading to the exclusion of Native Americans living on reservations, for example. The 1890 census was the first census in which census researchers were instructed to enumerate all Native Americans, taxed or not taxed, living on reservations or not. This change in method and measurement inaugurated in the 1890 census probably brought 200,000 or more Native Americans into census enumeration who were not enumerated in 1870 and 1880.[31]

There was something else that was new and different about the 1890 census: speed. Despite the significant increase in new questions, categories, and data collection, the 1890 census did not need more time for data processing and analysis. Indeed, the final results of the 1890 census were released and published after six years of analysis, two years less than the 1880 census. Many findings from the 1890 census were disseminated to the public before the complete publication of all findings in 1896. In 1892, a bulletin published by the superintendent of the census reported a change from previous census reports. Until 1890 "the country had a frontier of settlement, but at present the unsettled area has been so broken into isolated bodies of settlement that there can hardly be said [to] be a frontier line." As a result, from 1890 on, the measurement and analysis of a frontier line in western North America "cannot, therefore, any longer have a place in the census reports."[32] What made the frontier line disappear? Was it the growth of white settlers? Was it enumerating Native Americans? Was it a result not from changes in population patterns but from new measurements and methods? If the new questions, methods, measurements, Native American enumeration, and machine-based calculations of the 1890 census had been used in the 1880 census, would the 1880 census have broken the news about a disappearing frontier line? If the methods, measurements, machines, and Native American enumeration of the 1890 census had not been introduced until the 1900 census, would the

1900 census have been the first census to break the news to the nation that the frontier line had disappeared from the United States of America?

Whether these questions of new enumerations, machines, measurements, methods, and results ever crossed the mind of Frederick Jackson Turner is unknown. Turner did, however, immediately seize upon the value of the 1892 bulletin for his scholarship. Utilizing early findings from the big data of his era—the 1890 census report was a big data project significantly beyond the scale, scope, and ambition of its predecessors—Turner in 1893 proclaimed the disappearance of the frontier line and, therefore, the end of an era of American history. A brilliant act of scholarship from a brilliant person, Turner indelibly shaped the twentieth-century scholarship and sensibilities of American history by deftly deploying early and preliminary findings from (at that time) the largest and most comprehensive federal research project in the history of the nation. New data yielded new findings, and even more, big data yielded big findings. The new message from Turner—a line is gone, and an era is over—overshadowed the awareness and conversation about new methods and measurements. The nineteenth century witnessed the projection of all sorts of new lines across the North American frontier: telegraph lines; telephone lines; railroad lines; territorial, state, and county borders; isobars; time zones; latitude and longitude; a precise border with Mexico; streets, avenues, and other grids of white settlements; growing electrification and power lines; and the precise boundaries of Native American reservations. Amid a plethora of new lines on the frontier, an old frontier line was gone. The old frontier line disappeared as all the new lines emerged.

Turner looked at nature, the nation, and the North American frontier at the end of the nineteenth century. He saw a disappearing line, and heralded a new message that resonated for historians and for a vast array of Americans, from government officials to business and finance communities, white settlers, and the general public. Among the many who seized upon the new way in which Turner imagined nature and western North America were countless novelists and artists. They found the end of the frontier an endless font of inspiration for imagining a West that no longer is, but was. The message from Turner was exciting, new, inspirational, a bit romantic, and useful. Even better, no one had to stop his or her flow of imagination about nature and western North America. The American West was reimagined in ways that relegated the frontier to the past. At the same time, the new realization continued to allow Americans to imagine nature and the nation.

A newly discovered disappearing line spurred the national imaginary. Thinking back to the 1890 census, maybe all those new lines cast across the

North American frontier throughout the nineteenth century finally made the old line disappear from view. Twenty years or more of Native American surveillance by military telegraph combined with newly plotted lines of longitude and latitude made it easier to know where to find indigenous populations. Population numbers on any given Indian reservation were generally known by the army and by reservation agents. Maybe the 1890 census enumerated all Native Americans because, well, because it was easy by comparison with earlier censuses. No one any longer needed to travel endlessly across the American West on horseback, stagecoach, or—if all else failed—hitch a ride in an oxcart that stank of manure. If the railroad did not take you all the way there, the railroad got you close. All the new lines helped make it easier.

The 1893 message from Turner endured. It remains a mainstay of historical scholarship to this day. New interpretations, visions, and findings about the nation and nature emerged throughout the twentieth century. Ninety-nine years after Frederick Jackson Turner stood before a special meeting of the American Historical Association held in Chicago in 1893 in conjunction with the Columbian Exposition and presented a paper called "The Significance of the Frontier in American History,"[33] another brilliant person published a work based on preliminary findings about nature and the global environment. Those preliminary findings were based on new methods and new measurements. The message was drawn from the increasing capacity of scientists and researchers to newly deploy global electronic communication networks and monitor the environmental and climatological conditions and incipient changes for the planet. In the summer of 1992, Albert Arnold Gore Jr. published *Earth in the Balance: Ecology and the Human Spirit.*[34] The methods, measurements, data collection, and analysis presented by Gore in *Earth in the Balance* were, in retrospect, built on emerging scientific research about the global environment and climate but were also preliminary findings. Those findings, as well as the continuing methods, measurement, data collection, and analysis over the coming years proved accurate. Gore had a message about nature and the nation no less stunning than the 1893 message from Turner. But there was one huge difference between those two messages. The 1992 message from Gore proved over the years to be an inconvenient truth. The 1893 message from Turner proved over the years to be a convenient truth.

Inconvenient or convenient, Gore and Turner shared something in both their messages: common ground regarding the incredible significance, power, and value nature held over the imagination of Americans, past, present, and future. To imagine nature, first through the landscapes and ecosystems of western North America and now an imagination increasingly cast upon and even beyond the entire planet, is an enduring imagination that continues to resonate for Americans and for people around the world. When the telegraph wired into nature across the North American frontier during the latter half of the nineteenth century, that process brought a wealth of environmental and climatological data. It forged closer relations between the federal government and telegraph corporations. It launched tactics and strategies of surveillance and greater containment and control of Native Americans. It solidified mastery over the high ground, the signal flow, and the state secret. This all eventually coalesced in the secure command of electronic communication networks as part of the transformation of American expansion from continental to global parameters. Yet the centrality of nature to the American imagination remained, and continues to this day. Even if nature changed, everyone could still choose how to imagine the past, present, and future of nature. Hervey Johnston expressed this in a letter written home in April 1865. Hervey wrote as he sat somewhere atop the frontier high ground and gazed across the endless horizons.

> I don't know how it would seem in a more civilized country to look down from the mountains on a plain dotted with villages, and farm-houses, and marked up with a perfect network of roads, along which pours the busy population, and see every where, the inroads science and art are making upon nature, but here to look upon the unbroken solitude from the dizzy heights, inspires one with emotions such as none but a true lover of nature can appreciate.[35]

The challenge is not in finding the power or the motivation to imagine nature, but in finding the responsibility to imagine nature. This needs to be an enduring responsibility shaped by the growing evidence confirming the new message. What we have learned and need to practice after more than one hundred fifty years of being wired into nature is awareness, recognition, and responsibility. Some (but not all) messages about nature from the past proved incredibly convenient. The difficult and disturbing messages about the present and future of nature are now and will be for the most part inconvenient. Wiring into nature in the twenty-first century delivers a plethora of networks and abundant data confirming the deleterious impact of centuries of human activities upon the climate, environment, landscapes, ecosystems, and peoples of the planet. We need to imagine nature with disturbing truths. To do so is to practice responsible environmental citizenship and governance and transform imagination into action.

Notes

Introduction

1. Sergeant George H. Crane, Signal Service weather observer, Fort Gibson, Indian Territory, *1877 Annual Report of the Chief Signal Officer,* p. 35.

2. John, *Network Nation*; Hochfelder, *The Telegraph in America*; Gabler, *The American Telegrapher*; Starr, *The Creation of the Media*; Wolff, *Western Union*; Blondheim, *News over the Wires*; Downey, *Telegraph Messenger Boys*; Coe, *The Telegraph*; Phalen, *How the Telegraph Changed the World.*

3. "Electronic communication" is used herein as a shorthand phrase. Systems in use during the time of the study ranged across wire-based electromagnetic, electromechanical, and electronic systems.

4. Müller, *Wiring the World*; Winkler, *Nexus*; Wisnek and Pike, *Communication and Empire*; Nickles, *Under the Wire*; Schwoch, *The American Radio Industry*; Schiller, *Mass Communications and American Empire.*

5. For a review essay, see Müller, "From Cabling the Atlantic to Wiring the World."

6. *1879 Annual Report of the Chief Signal Officer*, p. 181.

7. *1887 Annual Report of the Chief Signal Officer*, p. 23.

8. Fleming, *Meteorology in America*, particularly chapter 6.

9. For an example of local barometer calibration in Madison, Wisconsin, see *1881 Annual Report of the Chief Signal Officer,* p. 52.

10. *Memoranda on the Work Accomplished by the Signal Service*, p. 8.

11. *1888 Annual Report of the Chief Signal Officer,* p. 11.

12. *1881 Annual Report of the Chief Signal Officer*, particularly pages 37, 42, 46, 67, and 91. For a compilation of press reports and testimonials from boards of trade, see

Memoranda on the Work Accomplished by the Signal Service. For a series of historical studies of weather observations at several Signal Service locations, see the Midwestern Regional Climate Center Climate Database Modernization Program, "Histories of Weather Observation at Selected Locations."

13. *St. Louis Republican* January 17, 1887, Box 8, William Babcock Hazen Papers (hereafter, Hazen Papers), National Museum of American History, Archives Center, Smithsonian Museum of American History, Washington, D.C.

14. "Extracts from Evidence before the Joint Commission upon Certain Bureaus referring to the Signal Corps," Washington 1886, pp. 55–56, Box 7, Folder Selected Federal Reports 1881–1886, Hazen Papers. Emphases in original.

15. Bates, *Lincoln in the Telegraph Office*; Wheeler, *Mr. Lincoln's T-Mails*; Wilson, *A Glimpse of the United States Military Telegraph Corps and of Abraham Lincoln.*

16. "One of the Old Guard," p. 225.

17. Coe, *The Telegraph,* pp. 128–29; "National Archives Celebrates the 145th Anniversary of Nevada Statehood."

18. Some of the results of the 1862 congressional elections were sent by telegraph to Secretary of State William Henry Seward. For Delaware, see Anson Stager note, Folder November 4, 1862; for Michigan, see Anson Stager note, Folder November 5, 1862; both in William Henry Seward Papers, University of Rochester, Rush Rees Library Rare Books, Special Collections, and Preservation Department, Rochester, NY.

19. Report from House Committee on Appropriations, "Signal-Service and Telegraph Companies," May 9, 1872, 42nd Congress, 2nd session, Report 69.

20. In the latter 1870s, Western Union and other commercial services were allowed to bill the government on a per-circuit basis, with one circuit generally defined as connectivity up to 500 miles.

21. Fleming, "Telegraphing the Weather," p. 170.

22. "History of the Signal Service," pp. 24–25.

23. *Regulations for the Operation and Maintenance of the United States Military Telegraph Lines,* p. 5.

24. For example, Cannon Ball, North Dakota, is geodetically located at 46° 14' 58" N/100° 38' 19" W. On the uses of telegraphy in plotting longitude and practicing geodesy, see Stachurski, *Longitude by Wire.*

25. Coe, *The Telegraph,* p. 137.

26. Thompson, *Wiring a Continent,* pp. 410–11.

27. For example, interconnectivity at Denison, Texas, between the military telegraph in Texas and the telegraph line built by the Missouri, Kansas, and Texas Railway Company in 1880 meant for the military telegraph system in that area "standard observatory time was obtained" (Ropp, *Hello—Anybody There?* p. 40).

28. Cronon, *Nature's Metropolis.*

29. The basis for my discussion of these terms are the definitions in the current online edition of the *Oxford English Dictionary*. I further place these definitions in narrative context with elaborations and examples.

30. For illustrated overviews of Farny, see Carter, *Henry Farny*, and Meyn, *Farny Paints the Far West*.

Chapter 1. Landscapes, Ecosystems, and Prevailing Westerlies

1. Lieutenant General Philip Sheridan to Senator George Spencer, March 23, 1878, cited in Ropp, *Hello—Anybody There?* p. 22. Sheridan wrote in support of a new telegraph line in Dakota Territory connecting Deadwood City, Black Hills, Fort Keogh, the mouth of the Tongue River, Yellowstone Valley, Fort Custer, the mouth of the Little Bighorn River, Bighorn Valley, and Fort Ellis. See also Senate Committee of Military Affairs report on military telegraph in Dakota and Montana Territories, April 9, 1878, Senate report 243, 45th Congress, second session, for supportive letters from Generals Sheridan, Sherman, and Crook.

2. Cronon, "The Trouble with Wilderness."

3. "Petition of Citizens of Minnesota Praying for a Grant of Land for the Establishment of a Telegraph from St. Paul, to the Telegraphic Lines South and East," May 21, 1850, Territorial Paper Series, State Territorial Papers Relating to Minnesota, Record Group 46, National Archives and Records Administration.

4. For example, see the 1855 treaty with the Blackfoot Nation in *Indian Affairs: Laws and Treaties*. The federal government, could, and often did, transfer or lease these rights, and similar clauses ensured rights for railroads. Similar language is found in the 1855 treaty with the Choctaws and Chickasaws, the 1893 (revised and proclaimed 1869) treaty with the western bands of Shoshone Indians, the 1855 treaty with Creeks and Seminoles, and the 1867 treaty with the Sisseton and Wahpeton Bands of Dakota or Sioux Indians.

5. Sunder, "St. Louis and the Early Telegraph, 1847–1857," pp. 248–58. The telegraph crossed the river at St. Louis by 1852 via a gutta-percha insulated underwater cable after failed local efforts to keep wire above and across the river. See also Scharf, *History of St. Louis City and County*, pp. 1427–28. Travelers reported lines across the Mississippi connecting Illinois and Iowa in 1851, probably near Rock Island and Davenport. The Iowa bank of the river also had some local telegraphic interconnectivity in the early 1850s. See Curtiss, *Western Portraiture and Emigrant's Guide*, p. 334.

6. Dillon, "The Arrival of the Telegraph in Texas."

7. Arrington, "The Deseret Telegraph." See also Bates, "The History of the Telegraph in California"; Sunder, "Telegraph Beginnings in Kansas"; West, ed., *The Essential West*, 78–99.

8. Ellis, "Lieutenant A. W. Greely's Report."

9. Ellis, "Lieutenant A. W. Greely's Report."

10. *A Manual of Military Telegraphy for the Signal Service*, p. 9. These instructions also were printed in *1872 Annual Report of the Chief Signal Officer*, pp. 611–13. Willow was to be avoided whenever possible.

11. Myer to Sheridan, July 20, 1878, and "General Instructions for the Construction of U. S. Military Telegraph Lines, Department of Dakota," no date, August 1878, both Box 64, A. W. Greely Papers, Manuscript Division, Library of Congress, Washington, D.C. (hereafter, Greely Papers). Emphasis in original.

12. Rue, "Pesh-Bi-Yalti Speaks," p. 255.

13. Ware, *The Indian War of 1864*, p. 70.

14. Greely to Myer, December 6, 1878, Box 64, Greely Papers.

15. Rue, "Words by Iron Wire," p. 69.

16. "A Manual of Military Telegraphy for the Signal Service," p. 43; *1872 Annual Report of the Chief Signal Officer*, pp. 611–13.

17. For an example of early stump rot from using green timber for telegraph poles around Ashland, Oregon, see *1888 Annual Report of the Chief Signal Officer,* p. 159.

18. Unrau, ed., *Tending the Talking Wire,* p. 134. The book is an edited collection of letters written by Hervey Johnson, an Ohio Quaker who enlisted in the Union Army and served on the Great Plains from 1863 to 1866.

19. "General Instructions for the Construction of U. S. Military Telegraph Lines, Department of Dakota," no date, August 1878, Box 64, Greely Papers.

20. Rue, "Words by Iron Wire," p. 87.

21. Grice, "History of Weather Observations, Fort Scully South Dakota 1866–1893," Midwestern Regional Climate Center Climate Database Modernization Program, September 2006, http://mrcc.sws.uiuc.edu/FORTS/histories/SD_Fort_Sully_Grice.pdf.

22. Ropp, *Hello—Anybody There?*, pp. 45–46.

23. Allin, "'A Mile Wide and an Inch Deep.'"

24. Ropp, *Hello—Anybody There?*, p. 33.

25. Coe, *The Telegraph,* p. 140.

26. *1879 Annual Report of the Chief Signal Officer*, p. 280.

27. On the April 1864 Idaho storm, see Unrau, ed., *Tending the Talking Wire,* p. 110; on the Julesburg incident, see Ware, *The Indian War of 1864*, p. 182.

28. Report of August 13, 1875, Box 68, Folder Military Papers: Reports 1875, Greely Papers. The report does not indicate whether these cedar trees were locally hewn or brought in from elsewhere.

29. Dillon, "The Arrival of the Telegraph in Texas."

30. Ellis, "Lieutenant A. W. Greely's Report," pp. 75–77.

31. On cutting and shipping trees from Puget Sound, see "Statement of the Origin, Organization, and Progress of the Russian-American Telegraph Western Union Extension, Collins' Overland Line, via Behring Strait and Asiatic Russia to Europe," May 1866, p. 14, Box 846; typescript "Pacific Communications," no date, but after 1900, Box 31, Folder Collins Overland Route (1 of 7); both in Western Union Telegraph Company Records, National Museum of American History Archives Center,

Smithsonian Museum of American History, Washington, D.C. (hereafter, Western Union Records). The "Pacific Communications" typescript reported "There was no suitable timber for poles close to either side of Behring Strait and some poles were bought and shipped from the Puget Sound mills."

32. Rue, "Words by Iron Wire," pp. 23–24.

33. Report of Richard J. Bush, May 16, 1866, Box 1, Folder Richard J. Bush, Western Union Telegraph Expedition Collection, 1865–1867, Smithsonian Institution Archives, Washington D.C. (hereafter, WUTE Papers).

34. Report of George Kennan, February 6, 1866, Box 1, Folder George Kennan, WUTE Papers. In the last decades of the nineteenth century, Kennan became a well-known author on Russian topics; see, for example, Kennan, *Tent Life in Siberia*. In 1863 Kennan was a Western Union telegraph operator in Cincinnati (Coe, *The Telegraph*, p. 70).

35. Report of George Kennan, April 9, 1866, Box 1, Folder George Kennan, WUTE Papers.

36. Report of Collins Macrae, June 4, 1867, Folder Collins Macrae; Report of D. C. Norton, Folder D. C. Norton, April 20, 1867; both in Box 1, WUTE Papers.

37. Unrau, ed., *Tending the Talking Wire*, p. 212.

38. Wolff, *Western Union*, p. 57.

39. *1875 Annual Report of the Chief Signal Officer*, p. 60.

40. *1877 Annual Report of the Chief Signal Officer*, p. 65.

41. *1881 Annual Report of the Chief Signal Officer*, p. 57.

42. Frank Pope to Charles Bulkley, May 5, 1866, report 3, in Reports of Western Union Telegraph Company explorations in western Canada, under Maj. Frank J. Pope 1865–1866, NAA MS 1682, National Anthropological Archives, Smithsonian Institution, Washington D.C. On the continuing challenges of maintaining military telegraph lines in the California and Oregon rainforests of the Pacific Northwest, see also *1887 Annual Report of the Chief Signal Officer*, p. 142.

43. *1878 Annual Report of the Chief Signal Officer*, p. 66.

44. *1884 Annual Report of the Chief Signal Officer*, p. 85.

45. "General Hazen's Reply to the Second Comptroller," 1886, p. 10, Box 7, William Babcock Hazen Papers, National Museum of American History, Archives Center, Smithsonian Museum of American History, Washington, D.C.

46. Replacement of wooden poles with iron masts in Arizona and New Mexico Territories took place throughout the 1880s; see *1885 Report of the Chief Signal Officer*, pp. 547–49.

47. *1887 Annual Report of the Chief Signal Officer*, p. 141.

48. *1888 Annual Report of the Chief Signal Officer*, p. 21.

49. Coutant, *The History of Wyoming*, p. 358.

50. *1888 Annual Report of the Chief Signal Officer*, p. 154.

51. *1886 Annual Report of the Chief Signal Officer*, pp. 180–83; Ropp, *Hello—Anybody There?*, p. 53.

52. *1883 Annual Report of the Chief Signal Officer*, p. 82; Ropp, *Hello—Anybody There?*, p. 33. Shooting an insulator installed on a telegraph or telephone pole was a popular activity, for target practice, shooting contests, and calibrating gunsights. The practice continues to this day in some areas. In 2015 the Durango Gun Club of Durango, Colorado, teamed up with the La Plata Electric Company to offer free gunsight calibration at the start of deer hunting season to discourage hunters from using insulators as calibration targets. See "Durango Gun Club Offers Free Gun Sighting-In to Prevent Insulator Damage," KOB-TV (Albuquerque), October 9, 2015.

53. Unrau, ed., *Tending the Talking Wire*, pp. 195–96.

54. *1883 Annual Report of the Chief Signal Officer*, p. 31. On monitoring locusts and other aspects of the Signal Service as a surveillance unit, see Fleming, "Telegraphing the Weather." On locusts, see Wagner, "Grasshoppered"; Berwanger, *The Rise of the Centennial State*, p. 11.

55. The dangers from rattlesnakes were recounted by Charles Brown, an assistant to Edward Creighton. See Mihelich and Potter, eds., *First Telegraph Line across the Continent*, p. 75. On skunks in the Platte Valley and at Fort Kearney, see Ware, *The Indian War of 1864*, p. 53, and on wolves stealing a sack of bacon, p. 234. On encounters with bears, see Unrau, ed., *Tending the Talking Wire*, pp. 291–92.

56. Rue, "Words by Iron Wire," p. 88.

57. Dillon, "The Arrival of the Telegraph in Texas," p. 210.

58. Reid, *The Telegraph in America*, pp. 504–505.

59. Coppedge and Shaw, "Effects of Horning and Rubbing Behavior by Bison."

60. Greenberg, *A Feathered River across the Sky*, particularly chapter 7. Native Americans had always harvested passenger pigeons whenever a roost alit nearby. Native Americans did not, however, exercise the capacity or wherewithal to slaughter passenger pigeons to extinction. Technologies of transportation and communication in the latter nineteenth century—railroads, Great Lakes maritime traffic, telegraphs, newspapers, and press services—served as a catalyst for hunters to achieve mass, rapid passenger pigeon slaughter to extinction and nearly achieved the same for bison.

61. James, *The First Scientific Exploration of Russian America*, pp. 94–95.

62. On reindeer herds and telegraph construction in Alaska, see Jackson, "Eleventh Annual Report on Introduction of Domestic Reindeer into Alaska, with Maps and Illustrations, 1901," particularly pp. 10, 23–24.

63. *1887 Annual Report of the Chief Signal Officer*, pp. 141–44.

64. For a study of one 1891 fort closure, see Wooster, "Fort Davis and the Close of a Military Frontier." On the impact of fort closures on telegraph line repairs, see Ropp, *Hello—Anybody There?*, p. 53.

65. Ropp, *Hello—Anybody There?*, pp. 58–64.

66. Henry, "The Passing of Signal Service, Weather Bureau Electric Telegraph and Cable Systems"; Eckel and Snow, "Coast Guard Landline Communications."

67. *1888 Annual Report of the Chief Signal Officer*, p. 19. Emphasis in original.

68. "History of the Signal Service with a Catalog of Publications, Instruments, and Stations," pp. 5–6, 25. Emphasis in original.

69. "The Proposed Partition of the Signal Service," 1880, in Box 64, Folder Orders, etc. December 1880, Greely Papers.

70. Wolff, *Western Union*, pp. 50–51.

71. Root and Hickman, "Pike's Peak Express Companies," p. 66.

72. Reid, *The Telegraph in America*, p. 498.

73. Gamble, "Early Reminiscences of the Telegraph"; Gamble, "Wiring a Continent."

74. Bates, "The History of the Telegraph in California"; Zornow, "Jeptha H. Wade in California"; Thompson, *Wiring a Continent*, pp. 344–47.

75. Rue, "Words by Iron Wire," pp. 16–18.

76. Sunder, "Arkansas' First 'Wonder Working Wire,'" p. 232.

77. Dillow, "Connecting Oregon"; Smith, "Communication Pioneering in Oregon."

78. Gamble, "Wiring a Continent," pp. 556–63. "Quaking asp" are aspen. See also Thompson, *Wiring a Continent*, pp. 363–64.

79. Arrington, "The Deseret Telegraph."

80. Gamble, "Wiring a Continent," p. 562.

81. Coutant, *The History of Wyoming*, p. 637.

82. Sandbo, "Westward to Europe," undated manuscript, Alaska State Library and Archives, Juneau, AK. Investors in Denver refused to pay the $20,000 needed to have the Transcontinental Telegraph run through Denver.

83. Sunder, "Telegraph Beginnings in Kansas," pp. 32–38.

84. Sandbo, "Westward to Europe."

85. Sunder, "Arkansas' First 'Wonder Working Wire.'"

86. The congressional bill was the Pacific Telegraph Act of 1860, "An Act to Facilitate Communication between the Atlantic and Pacific States by Electric Telegraph," 36th Congress, 1st session, June 16, 1860. See Thompson, *Wiring a Continent*, pp. 349–55 and 515–17.

87. Coe, *The Telegraph*, pp. 38–39.

88. Root and Hickman, "Pike's Peak Express Companies," pp. 40–57.

89. Thompson, *Wiring a Continent*, pp. 361–68.

90. *Omaha Times*, September 16, 1858, cited in Nielson, "Edward Creighton and the Pacific Telegraph," pp. 61–74.

91. Sunder, "The Early Telegraph in Rural Missouri, 1847–1859."

92. Root and Hickman, "Pike's Peak Express Companies," p. 38; Thompson, *Wiring a Continent*, pp. 333–34.

93. Reid, *The Telegraph in America*, p. 495.

94. Ware, *The Indian War of 1864*, p. 9.

95. Mihelich and Potter, eds., *First Telegraph Line across the Continent*, pp. 20–21; Wolff, *Western Union*, p. 57.

96. On the uses of trees for military installations, see Unrau, ed., *Tending the Talking Wire*, p. 111; on Cottonwood Springs, see Ware, *The Indian War of 1864*, p. 61. Ware

recounted that at Fort McPherson the furniture "was all made out of beautiful red cedar."

97. On stagecoach seats, see Ware, *The Indian War of 1864*, p. 27. On well digging, see Mihelich and Potter, eds., *First Telegraph Line across the Continent,* pp. 55–56.

98. Root and Hickman, "Pike's Peak Express Companies," p. 68.

99. Jeptha Wade, October 29, 1861, cited in Unrah, ed., *Tending the Talking Wire,* p. 50.

100. Root and Hickman, "Pike's Peak Express Companies," p. 69.

101. *St. Joseph* (Missouri) *Weekly Free Democrat*, October 27, 1860, cited in Root and Hickman, "Pike's Peak Express Companies," p. 70.

102. Army troops guarding the Transcontinental Telegraph during and after the Civil War were among the many that used the telegraph for money transfers. Hervey Johnson wrote his parents on April 11, 1865, that if he had spare money on the next payday he would "deposit it in the hands of the telegraph company" and the telegraph company would then deliver the money from Hervey to his family at home. Unrau, ed., *Tending the Talking Wire,* p. 237. On telegraph traffic and the Comstock Lode, see Coe, *The Telegraph,* p. 90.

103. Nielson, "Edward Creighton and the Pacific Telegraph," p. 73.

104. Thompson, *Wiring a Continent,* pp. 363–68.

105. Howe, *What Hath God Wrought,* pp. 813–18.

106. Thompson, *Wiring a Continent,* p. 343.

107. Scheips, "Albert James Myer, an Army Doctor in Texas, 1854–1847."

Chapter 2. Storms Moving in a Ring of Fire

1. Nielson, "Edward Creighton and the Pacific Telegraph," p. 61.

2. Lockwood and Lockwood, *The Siege of Washington*, pp. 47–49.

3. Cooling, "Defending Washington," p. 318.

4. Hendrickson, "Defending Washington," p. 46; Billings, "Military Activities in Washington," p. 127.

5. Lockwood and Lockwood, *The Siege of Washington,* pp. 173–74; Hochfelder, *The Telegraph in America,* p. 8.

6. Kielbowicz, "The Telegraph, Censorship, and Politics," p. 98.

7. Lockwood and Lockwood, *The Siege of Washington,* p. 229.

8. Spencer, *History of the United States*, p. 22: "It was therefore resolved to make a forward movement into Virginia. . . . [I]mmediate possession was taken of Arlington Heights and of Alexandria. At this latter place, Colonel Ellsworth . . . arrived by water, very early in the morning of the 24th of May. His first impulse was to destroy the railroad communication and seize upon the telegraph office . . . but as he was on his way to the telegraph office, he espied flying from the Marshall House, a second class hotel, a Confederate flag. . . . (Ellsworth cut down the flag) and having wrapped it around his body, was coming down the stairs. The proprietor of the house, a man

by the name of Jackson, met him, and seeing what he had done, fired into his bosom. Jackson immediately thereafter was killed by one of the Zouaves in company."

9. Kielbowicz, "The Telegraph, Censorship, and Politics," p. 96.

10. Schwartzlose, *The Nation's Newsbrokers*, pp. 241–42.

11. Hochfelder, *The Telegraph in America*, p. 7.

12. Sunder, "Arkansas' First 'Wonder Working Wire,'" p. 242.

13. Wolff, *Western Union*, pp. 61–64.

14. Thompson, *Wiring a Continent*, p. 373.

15. In April 1861, General George McClellan asked Anson Stager to oversee telegraph security in the Department of the Ohio. Hochfelder, *The Telegraph in America*, pp. 8–9; see also Reid, *The Telegraph in America*, p. 483.

16. Peters, *Seven Trails West*, p. 187.

17. Coe, *The Telegraph*, p. 43.

18. Unless otherwise indicated, this paragraph on Myer and his transition from the Navajo campaign to the Civil War is based on Raines, *Getting the Message Through*, pp. 5–13.

19. Brown, *The Signal Corps*, chapter 3, discusses the Red Hill location. Red Hill is at approximately R Street NW between Wisconsin Avenue and 32nd Street NW. See Fletcher, "The Signal Camp of Instruction" at the *Glover Park History* website.

20. Brown, *The Signal Corps*, p. 171.

21. Ibid., p. 63.

22. Kielbowicz, "The Telegraph, Censorship, and Politics," pp. 95–118.

23. Schwartzlose, *The Nation's Newsbrokers*, pp. 65–72. The nearest place on the network varied, in part because of the expansion of the telegraph networks, in part because of technical failures on existing networks and at telegraph stations. One example of this condition was news about the end of the war reported by telegraph with a dateline of Conway, South Carolina: hundreds and hundreds of miles from the front.

24. Scheips, "Union Signal Communications," p. 399. The Crimean War is generally seen as the first major conflict utilizing the telegraph as a military asset. For a discussion in the context of telegraphy and the Ottoman Empire, see Bektas, "The Sultan's Messenger, 1847–1880."

25. Cox, "The Defenses of Washington," p. 135.

26. Cooling, "Defending Washington," p. 316.

27. Plum, *The Military Telegraph*, is a detailed account of the USMT. See also Johnson, "Reminiscences of the Honorable Edwin M. Stanton"; Scheips, "Union Signal Communications."

28. Joshua Wolff described the telegraph industry as "in effect, [a] privatized part of the military" (Wolff, *Western Union*, p. 49).

29. Greely, "The Military-Telegraph Service," in Miller and Lerner, eds., *The Photographic History of the Civil War*.

30. Quartermaster General Montgomery Meigs "insisted" on military appointments for some USMT officers "in order to streamline ordering of telegraph supplies and equipment" (Hochfelder, *The Telegraph in America,* p. 9).

31. For a brief summary, see Pfeiffer, "Records Relating to the U. S. Military Railroads."

32. Thompson, *Wiring a Continent,* pp. 385–88.

33. Brown, *The Signal Corps,* p. 178.

34. Greely, "The Military-Telegraph Service."

35. Grant, *Personal Memoirs of U.S. Grant,* chapter 51. Thanks to Robert Hariman for alerting me to this passage. Hariman astutely observed that Grant was, in a way, describing what is known as a "hot spot" in contemporary internet use.

36. Hochfelder, *The Telegraph in America,* p. 20, on cotton; on gold speculation, including by Anson Stager, see idem, p. 104.

37. Pound, "The Military Telegraph and the Civil War"; quoted passages at pp. 188, 196.

38. Brown, *The Signal Corps,* p. 174.

39. Myer, "Report of the Chief Signal Officer Army of the Potomac Previous to and During the Campaign on the Peninsula, Virginia, Rendered to Major General George B. McClellan," October 21, 1862, p. 5, in Records of the Signal Corps, Record Group (hereafter RG) 111, entry 9, Box 1, National Archives and Records Administration (hereafter, NARA).

40. Plum, *The Military Telegraph,* 2:88.

41. Hochfelder, *The Telegraph in America,* p. 11.

42. Thompson, *Wiring a Continent,* p. 388.

43. Anson Stager found his USMT responsibilities demanding, but he also commented in 1862 that he would remain in his military position until he could step down "without detriment to the Telegraph interests of the country" (Stager, quoted in Wolff, *Western Union,* p. 67).

44. Scheips, "Union Signal Communications." The system was the Beardslee Telegraph, advantageous in that it used a hand-turned magneto, eliminating the need for batteries; problematic in its short range of less than ten miles, slow message transmission, and need for a dedicated circuit.

45. Johnson, "Reminiscences of the Honorable Edwin M. Stanton," pp. 72–74.

46. Blondheim, "Public Sentiment Is Everything"; Hochfelder, *The Telegraph in America,* pp. 86–91; Kielbowicz, "The Telegraph, Censorship, and Politics"; Schwartzlose, *The Nation's Newsbrokers,* pp. 242–54. See also Plum, *The Military Telegraph,* 2:147–48, for a brief mention of Stafford G. Lynch as War Department Telegraph Office Censor of the Press for part of 1863.

47. Hochfelder, *The Telegraph in America,* pp. 26–27; Wolff, *Western Union,* pp. 82–83.

48. On the rebuilding of lines, operations in the Confederacy throughout 1865, and the transition to private ownership, see Plum, *The Military Telegraph,* 2:340–50.

49. Peters, *Seven Trails West*, p. 179.

50. Johnson, "Reminiscences of the Honorable Edwin M. Stanton," p. 72.

51. *1864 Annual Report of the Signal Officer of the Army*, October 31, 1864, p. 8, stated that orders of November 10, 1863, directed Signal Corps telegraph trains and equipment "to be turned over to the United States Military Telegraph Company; which was promptly done. But when the emergency has seemed to demand it, some of the Department Commanders have directed them to be returned to the Corps for temporary use." This indicates that some flexibility remained for officers in the field in the deployment of the telegraph. The 1864 report was filed by Acting Signal Officer Lieutenant Colonel William J. L. Nicodemus. For the report, see RG 111, entry 33, Box 1, NARA.

52. Plum, *The Military Telegraph*, 2:102, has a copy of the order relieving Myer of command.

53. Brown, *The Signal Corps*, pp. 592–94.

54. Plum, *The Military Telegraph*, 2:48–49; Curtis passage at p. 215.

55. Stager to Stanton, November 1, 1863, "Report of the Operation and Conditions of the U.S. Military Telegraph for the Fiscal Year Ending June 30, 1863," online at http://www.civilwarsignals.org/pages/signal/signalpages/or5.html#top.

56. Wolff, *Western Union*, p. 68.

57. "Account of the stock owned by Western Union Telegraph Company in other lines out of its territory," October 20, 1863, Box 1, Folder 20, Hiram Sibley Family Papers, University of Rochester Rush Rees Library Rare Books, Special Collections, and Preservation Department, Rochester, NY.

58. McDermott, *Circle of Fire*, p. 9.

59. Jepsen, "The Telegraph Comes to Colorado"; Berwanger, *The Rise of the Centennial State*, p. 12.

60. Ware, *The Indian War of 1864*, p. 32.

61. Jepsen, "The Telegraph Comes to Colorado," p. 5.

62. Unrau, ed., *Tending the Talking Wire*, pp. 87, 114 on deserters; p. 147 on tying people to a pole.

63. Jepsen, "The Telegraph Comes to Colorado"; Peters, *Seven Trails West*, pp. 173–94; Gamble, "Wiring a Continent," pp. 556–63.

64. Thompson, *Wiring a Continent*, p. 365.

65. This story of intentionally shocking indigenous peoples is often attributed to Edward Creighton. For one example from a diary, see Mihelich and Potter, eds., *First Telegraph Line across the Continent*, pp. 90–91, a collection of entries from Charles Brown's diary. Brown added that later that day, when telegraph wire was strung over a Native American encampment, "in hot haste the squaws *and also the bucks* tore down their lodges and decamped." Emphasis in original.

66. Ware, *The Indian War of 1864*, p. 387.

67. Berwanger, *The Rise of the Centennial State*, p. 7; "Report of the John Evans Study Committee," p. 18.

68. Berwanger, *The Rise of the Centennial State*, p. 18; Convery, "John M. Chivington."

69. "Report of the John Evans Study Committee," pp. 56, 61, 64.

70. Unrau, ed., *Tending the Talking Wire*, p. 178.

71. Jepsen, "The Telegraph Comes to Colorado," p. 15.

72. "Report of the John Evans Study Committee," p. 73.

73. McDermott, *Circle of Fire*, p. 12.

74. Berwanger, *The Rise of the Centennial State*, pp. 19–20. Mining communities in Colorado spoke out in opposition to Sand Creek and the "indiscriminate massacre of all Indians." On Colorado newspaper coverage and editorials about the Sand Creek Massacre and its immediate aftermath, see Reilly, *The Frontier Newspapers*, pp. 17–30.

75. Ware, *The Indian War of 1864*, p. 324.

76. Jepsen, "The Telegraph Comes to Colorado," p. 8; Ware, *The Indian War of 1864*, p. 320–22.

77. For Mud Springs, see Mihelich and Potter, eds., *First Telegraph Line across the Continent*, p. 110n68; for the twenty-mile string of attacks near Julesburg, see Berwanger, *The Rise of the Centennial State*, p. 23.

78. Unrau, ed., *Tending the Talking Wire*, pp. 187–88.

79. McDermott, *Circle of Fire*, p. 67.

80. Ibid., p. 23.

81. Resolution of the Legislative Assembly, Territory of Nebraska, February 13, 1865.

82. Berwanger, *The Rise of the Centennial State*, p. 24.

83. McDermott, *Circle of Fire*, p. 142.

84. Ibid., p. 84.

85. Ibid., p. 96; see also "Diary of Jake Pennock."

86. Ibid., pp. 158–60; see also "Diary of Jake Pennock."

87. Unrau, ed., *Tending the Talking Wire*, pp. 260–80.

88. McDermott, *Circle of Fire*, pp. 3–9.

89. Ware, *The Indian War of 1864*, pp. 364–72.

90. Jepsen, "The Telegraph Comes to Colorado," p. 18.

91. Ware, *The Indian War of 1864*, pp. 354–57; "The fire rolled on, leaving in its train only blackness and desolation."

92. On the fire, see also McDermott, *Circle of Fire*, pp. 27–29. McDermott observed the "main objective in setting the fire was to deprive raiders of forage for their ponies and drive all game out of the region, reducing their subsistence."

93. Ware, *The Indian War of 1864*, p. 357.

94. Hamblin, *Arming Mother Nature*.

95. Standage, *The Victorian Internet*.

96. Lukasik, "Why the Arpanet Was Built," pp. 4–21.

97. "Risk areas" refers to an approximation of square miles potentially at risk, and not a calculation of square miles that suffered scorched-earth tactics. The Great

Prairie Fire took place in areas largely unpopulated by white settlers; accounts of the event are scarce. Furthermore, it was not a forest fire but a grassland fire, making comparisons to major forest fires challenging. Some areas of the fire spread as far south as the Texas Panhandle. Even with these approximations and scant accounts, it is likely that this was the largest intentionally set fire ever lit on the North American continent, and it likely spread over a risk area larger than any known forest fire in North America. Cottonwood Canyon, where Mitchell gave the telegraph orders to ignite the fire, is the same general vicinity as the present-day city of North Platte, Nebraska.

98. National Park Service, *Santa Fe Trail Map Timeline 1866–1873*.

99. Ware, *The Indian War of 1864*, pp. 383–84.

100. McDermott, *Circle of Fire*, p. 33.

101. Unrau, ed., *Tending the Talking Wire*, pp. 246–48.

102. Jepsen, "The Telegraph Comes to Colorado," p. 14; Reid, *The Telegraph in America*, p. 494. Reid added "The Indians were generally friendly. Some of the Indian chiefs still send Mr. Sibley gifts of live deer and antelope as tokens of their good will" (p. 494). Hiram Sibley took an approach centered on friendship and understanding with indigenous peoples, an approach not always followed by other telegraph builders and operators.

103. When I was very young—no more than three or four years old—I sometimes visited an elderly man in our neighborhood, then in his upper nineties. He told me when he was my age, he vividly remembered the assassination of Abraham Lincoln, because as soon as the first news arrived in town, he immediately went to the local telegraph station where he and many others spent all day as reports came in.

104. Jepsen, "The Telegraph Comes to Colorado," pp. 14–15.

105. For a critique of this attitude toward the telegraph and supposition of white civilization's monopoly of technical knowledge in a more global context, see Müller, *Wiring the World*, pp. 106–18.

106. Ware, *The Indian War of 1864*, p. 20. See also p. 6 for a report of a white settler woman who was at first unnerved by this behavior of Native Americans looking in her window.

107. One early realization in the British efforts to decrypt German Enigma machines during World War II was based on German signal traffic: because the German military always sent out a message to all troops and units at 6:00 a.m. every day, the British inferred, although Enigma was not yet decrypted by British codebreakers, that that message was most likely a weather forecast. See Singh, *The Code Book*, pp. 170–78.

108. McDermott, *Circle of Fire*, p. 105.

109. "Dumpster diving" or going through the unsecured trash of a telecommunication business remains one activity of hackers today. It was a major factor for the success of "phone phreaks" in surreptitiously accessing telephone service, particularly before the rise of the paper shredding machine. Fully securing discarded electronic equipment, software, storage devices, and similar electronic communication waste

technologies and data from rummaging by local populations has proven incredibly difficult. George Bent, a Cheyenne who also was at one point in the Confederate Army, is known to have fought with and assisted Great Plains warriors in several ways, including explaining canned food and paper currency. For one example of guidance from Bent, see Berwanger, *The Rise of the Centennial State,* p. 25.

Chapter 3. Changes in the Forecast

1. *1874 Annual Report of the Chief Signal Officer*, p. 518.
2. John, *Network Nation*, p. 116.
3. An Act to Increase and Fix the Military Peace Establishment of the United States, 39th Congress, 1st session, July 28, 1866. See also General Order no. 56, War Department, Adjutant General's Office, 1866.
4. Unless otherwise indicated, this discussion of George Maynard in this chapter is based on the following: Van Orsdel, "History of the Telephone System in the District of Columbia"; Loomis, "The Telephone Comes to Washington"; "Tragedy of April 14, 1865, as recalled by Eight Survivors of the U. S. Military Telegraph Corps," *Milwaukee Sentinel*, April 11, 1915. On the national dissemination of news on the assassination of Lincoln, see also Brown, *Knowledge Is Power,* pp. 258–67. Late in his career Maynard was the original organizer and curator of what became the electrical equipment, telephone, and telegraph collections of the Smithsonian. Maynard's notes, correspondence, papers, and manuscripts, consulted by Loomis (for the article cited here) while those papers were still held by Maynard descendants, have now been consigned to auction houses, as some online searches for Maynard found an auction house that is selling (or has sold) portions of the Maynard Papers.
5. Guiteau and Maynard were distantly related. See Millard, *Destiny of the Republic,* chapter 10; Yanoff, *The Second Mourning.* For testimony by Maynard at the Guiteau trial, see *Report of the Proceedings in the Case of the United States vs. Charles J. Guiteau,* pp. 220–23; 877–82. Maynard must have been a sterling witness for the prosecution, for he had previously installed the telephone system in the Supreme Court offices. See Loomis, "The Telephone Comes to Washington."
6. "History of the Signal Service"; Loomis, "The Telephone Comes to Washington"; *Memoranda on the Work Accomplished by the Signal Service; Observation and Report of Storms*; Miller, "New Light on the Beginnings of the Weather Bureau." See also Fleming, "Storms, Strikes, and Surveillance"; Raines, *Getting the Message Through*, chapter 2.
7. On the development of meteorology in the nineteenth century up to the 1870 inauguration of the Signal Service, see Fleming, *Meteorology in America.* Fleming's masterful study includes an excellent chapter on the growing importance of the telegraph to American weather forecasting in the 1850s and 1860s (chapter 7, pp. 141–62).
8. A note on terminology for "Signal Corps" and "Signal Service": generally, Signal Corps refers to the military branch of the army founded in 1859 and centrally tasked

with signal communication for the army, regardless of the various technologies used for signaling (flags, torches, flares, telegraph, radio, and so forth). Signal Service refers to the specific branch within the Signal Corps created in 1870 to carry out the congressional mandate for national weather data collection and analysis, including weather forecasts. The Signal Service performed that task until 1890–91, at which point the weather project was transferred to the Department of Agriculture coincident with the creation of the National Weather Service (NWS), now the National Oceanic and Atmospheric Administration (NOAA). Despite the clear delineations, there was nevertheless a blurring of the Signal Corps and the Signal Service in many ways. Both entities shared the same military telegraph facilities, both entities could get involved in surveillance against Native Americans as well as other nationwide surveillance activities, and both reported to the same general command staff. Furthermore, government reports and other archival documents of the era often use the term "Signal Office" to describe either the Signal Corps or the Signal Service, or just as often, both the Signal Corps and the Signal Service. James Fleming is among those scholars who often use the term "Signal Office" in their research, and his use of "Signal Office" is a sensible choice that is faithful to the often conjoined and overlapping missions of the Signal Corps and the Signal Service. Eric Miller found the following terms used in documents of the era: Signal Service, Signal Corps, Signal Office, Signal Detachment, Signal Force, and Signal Bureau (Miller, "New Light on the Beginnings of the Weather Bureau," p. 69). I largely stick to Signal Corps and Signal Service, although additional terms appear in some notes.

9. "History of the Signal Service," p. 4.

10. *1867 Annual Report of the Chief Signal Officer*, p. 4. See also Howgate and Greely to Myer, March 5, 1869, RG 111, Entry 7, Correspondence Relating to Cipher Systems, 1868–1870, Box 1, National Archives and Records Administration (hereafter, NARA). The Alberti disk was a handheld device that determined the alphabetic shifting of letters, for example a=c, b=d, c=e, and so forth. Dating to the 1700s and itself an advancement on ancient encryption techniques such as a Caesar Shift, various complexities and multiple letter shifts had been introduced into the Alberti disks used by the Union Army during the war. Anson Stager also created some ciphers and codes for the USMT. The resultant Civil War Union ciphers were secure in the short term but could be decrypted eventually through frequency analysis (counting the frequency of individual letters in a cipher to predict more commonly used letters in plaintext English words such as e, a, n, s, l, or t). The television game show *Wheel of Fortune* is in part based on the contestant's abilities to apply plaintext frequency analysis with common letter combinations and discern individual words.

11. *1870 Annual Report of the Chief Signal Officer,* p. 6.

12. Fleming, *Meteorology in America,* particularly chapter 4; Monmonier, *Air Apparent,* particularly chapter 3.

13. John, *Network Nation,* pp. 123–24.

14. *1870 Annual Report of the Chief Signal Officer*, p. 104.

15. Ibid., p. 105.
16. Ibid., p. 107.
17. Ibid., p. 108.
18. Ibid., pp. 136–38.
19. "A Chronological Outline of the History of Meteorology in the United States."
20. *1871 Annual Report of the Chief Signal Officer*, p. 262.
21. "Brief of Papers on File in the Office of the Chief Signal Officer of the Army Relative to the Telegraph Service of the Western Union Telegraph Company," various entries from 1870 and 1871, Reel 2, Albert James Myer Papers, Library of Congress Manuscript Division, Washington, D.C. (hereafter, Myer Papers).
22. *1871 Annual Report of the Chief Signal Officer*, p. 266.
23. Ibid., p. 267.
24. Ibid., p. 271.
25. Western Union president William Orton, in Report from House Committee on Appropriations, "Signal-Service and Telegraph Companies," May 9, 1872, 42nd Congress, 2nd session, Report 69, p. 7 (hereafter, *1872 House Appropriations Signal Service Report*). The committee chair was future president James Garfield. Additional congressional documents relevant to this 1872 discussion of Signal Service ciphers and commercial telegraph traffic were Senate Committee on Commerce, June 1, 1872, 42nd Congress, 2nd session, Report 223. Unless otherwise indicated, the discussion herein is based on the House Appropriations Committee report.
26. On the development of reading and relaying telegraph messages by sound, see Coe, *The Telegraph,* pp. 66–70.
27. Orton, in *1872 House Appropriations Signal Service Report*, p. 31.
28. See, for example, the 1875 Signal Service Telegraph Cipher codebook, RG 111, Entry 40, NARA, p. 9, for cipher words representing atmospheric pressure measurements from 30.00 to 30.59.
29. Orton, *1872 House Appropriations Signal Service Report*, p. 8.
30. Ibid., p. 30.
31. Headquarters of the Department of the Missouri, Circular 13, Ft. Leavenworth, Kansas, April 24, 1883.
32. Hawes, "The Signal Corps," pp. 68–76.
33. The National Telegraph Act of 1866 empowered the postmaster general to negotiate and set federal government rates for telegraph services with the commercial telegraph networks. For a summary, see John, *Network Nation*, pp. 116–22.
34. Orton, *1872 House Appropriations Signal Service Report*, pp. 9–10.
35. Fleming, "Telegraphing the Weather," p. 163.
36. "All such persons as have been or shall be designated and employed [by the Chief Signal Officer] for the taking of meteorological observations, or for preparing or presenting telegraphic communications for transmission, or for transmitting the same, are hereby recognized and appointed as agents of the War Department, for those purposes, and are hereby authorized and directed by and in behalf of said Department to offer any telegraph company in the United States for transmission

any and all such telegraphic communications as they may be required by the Chief Signal Officer to make, and to request the transmission thereof by such company or companies, at such times and in such places as may be directed by the Chief Signal Officer." Order issued by the secretary of war on June 27, 1872, found in the compendium "Legislation and Orders Related to the Signal Service, U.S. Army from 1860 to 1873," Reel 1, Myer Papers.

37. Pamphlet by Myer, no date but 1875, Reel 1, Myer Papers.

38. Myer was promoted to brigadier general in 1866, an outcome of the 1866 army reorganization. Prior to World War I, this reorganizational plan transformed the Signal Corps commander into a plum job for career army officers, many of whom lobbied for the appointment when it became vacant. Various army regulations and organizational schemata from about 1866 to 1917 restricted the total number of allowable appointments to the rank of general. An additional prize for the Signal Corps commander was the required posting in Washington with its access, visibility, and social life.

39. For a recent review of major works and themes in media archeology, see the review essay by Natale, "Understanding Media Archeology."

40. Soll, *Information Master.* I am indebted to Richard John for alerting me to this brilliant book.

41. For a study that takes up the issue of government data collection after 1870, see Cortada, *All the Facts.*

42. Fleming, "Telegraphing the Weather," pp. 167–69. A gender issue casts a dark and disturbing shadow across this incident, hinting at a genealogy of men stalking women via electronic communication networks and social media.

43. Unterman, *Uncle Sam's Policemen.*

44. Fleming, "Telegraphing the Weather," p. 165. Fleming reproduces some telegrams, all sent to "CSO" or Chief Signal Officer Albert Myer.

45. Bruce, *1877: Year of Violence*, pp. 212–19, 274, 291; Hochfelder, *The Telegraph in America*, pp. 93–94.

46. *New York Times*, "Action of the Government," July 26, 1877.

47. *New York Times*, "Signal Service Proficiency," August 10, 1877.

48. *New York Times*, "Hayes's Administration," March 2, 1881.

49. Grant quotes and discussion of the 1877 Railroad Strikes from Gelderman, *A Free Man of Color*, p. 122.

50. Clark to Blaine, May 9, 1874, "Telegraph-Offices in the Corridors of the Capitol," House of Representatives, 43rd Congress, 1st session, misc. doc. 269.

51. The reliability of the "Willard's men" might beg credulity, in that Willard's Hotel was the most important Washington locale for the gathering of lobbyists, congressional representatives and senators, other government workers, and office-seekers. The credibility of telegraph operators was rarely questioned.

52. Robert T. Lincoln, secretary of war, to Senate, June 25, 1884, report on underground telegraphic cables laid between the Capitol and executive departments, 48th Congress, 1st session, ex. doc. 197; letters about "underground telegraph and

telephone wires" from the secretary of war and the commissioners of the District of Columbia to Senate, April 12, 1888, 50th Congress, 1st session, ex. doc. 153; report of the architect of the Capitol to the secretary of the interior about November 6, 1898, gas explosion in the Capitol and subsequent repairs, July 1, 1899.

53. Letters about "underground telegraph and telephone wires" from the secretary of war and the commissioners of the District of Columbia to Senate, April 12, 1888, 50th Congress, 1st session, ex. doc. 153; Gilmore, "Washington Goes Wireless." On problems with local Washington police telephone call boxes, see Appropriation for Telegraph and Telephone Service, District of Columbia, March 25, 1897, H.R. Document 11, 55th Congress, 1st session. Second Precinct Lt. D. H. Teeple reported that in the second police district, "It is only at rare intervals that any intelligent conversation can be carried on (and for that matter, any conversation at all) between the officers at the patrol boxes on their respective beats and the officer at the station house."

54. Unless otherwise noted, this discussion of Maynard and the Washington telephone is from Bolce, "The Telephone at the White House"; Loomis, "The Telephone Comes to Washington"; Loomis, "The White House Telephone and Crisis Management"; and Van Orsdel, "History of the Telephone System in the District of Columbia."

55. A directory facsimile is in Loomis, "The Telephone Comes to Washington."

56. Heritage Auctions, Archive of Agreements and Reports for Early Phone Lines in Washington, D.C., portion of George Maynard papers, online at http://historical.ha.com/itm/miscellaneous/-bell-telephone-archive-of-agreements-and-reports-for-early-phone-lines-in-washington-d-c/a/6084–34168.s.

57. Gelderman, *A Free Man of Color*, pp. 12–15, 46–47, 89, 99–117. See also Wynes, "James Wormley."

58. Scott was a Wormley boarder for many years and regularly ate catered food from Wormley; he was particularly fond of terrapin (Gelderman, *A Free Man of Color*, pp. 14–15).

59. On Hazen's home, see memorial book "Press Sketches of the Life and Services of General W. B. Hazen, Chief Signal Officer, U.S.A. private edition, 1887," Box 8, William Babcock Hazen Papers, National Museum of American History, Archives Center, Smithsonian Museum of American History, Washington, D.C.; on Signal Corps officers residing near Wormley businesses, see "Officers of the United States Army on Duty or Residing in the District of Columbia," January 1, 1875, (and 1876, 1877, and 1879), Boxes 63 and 64, Adolphus Washington Greely Papers, Manuscript Division, Library of Congress, Washington, D.C.

60. *1877 Annual Report of the Chief Signal Officer*, p. 8.

61. See "The Telegraphic Historical Society of North America." Benjamin Montgomery also was a founding member of the historical society. The Smithsonian agreed in 1894 to "provide a place in the National Museum for telegraphic relics" collected and donated by the Telegraphic Historical Society (p. 26).

62. Henson, "Objects of Curious Research," pp. 249–69.

63. *DAR Handbook*, p. 35.

64. Hochfelder, *The Telegraph in America,* pp. 58–62.

65. *1884 Annual Report of the Chief Signal Officer,* p. 18.

66. In 1877, Western Union signed a five-year contract with Thomas Edison for technology development, Edison already having developed the stock ticker and a quadruplex system. As a teenager, Edison was a telegraph operator with the USMT. Coe, *The Telegraph,* pp. 118–19.

Chapter 4. Dreams of a Boreal Empire, Nightmares of a Polar Vortex

1. William Henry Seward to Zachariah Chandler, May 14, 1864, "Communication of Hon. William H. Seward upon the Subject of an Intercontinental Telegraph Connecting the Eastern and Western Hemispheres by Way of Behring's Strait"; also cited in Vivier, "The Collins Overland Line and American Continentalism," p. 247.

2. The phrase "mutually beneficial result" is from Postinov and Falk, *Exploring and Mapping Alaska,* p. 405.

3. Vasile, "The Early Career of Robert Kennicott." Lapham (Madison and Milwaukee), Hoy (Racine), and Kirtland (Cleveland) also were located in the Midwest.

4. Peck, "New Records of the Lincoln-Douglas Debate," pp. 55–59. On John Kennicott, see Ernst, "John A. Kennicott of the Grove."

5. Kennicott, "Description of a New Snake." Kirtland's snake (*Clonophis kirtlandi*) is native to the American Midwest, roughly from the southern shores of the Great Lakes across Ohio, Indiana, and Illinois south to Kentucky. It is classified as either endangered or threatened throughout its range.

6. For biographical and background information on Robert Kennicott, see Binnema, *Enlightened Zeal*; Clark and Forbes, "Science in Chicago"; Coates, "The Kennicott Network"; Deignan, "The HBC and the Smithsonian"; Fitzhugh, "Origins of Museum Anthropology at the Smithsonian"; Foster, "Robert Kennicott"; Hendrickson, "Robert Kennicott"; James, *The First Scientific Exploration of Russian America*; Laubacher, "The 'Mandarin-Missionary' Strategy"; Lindsay, ed., *The Modern Beginnings of Subarctic Ornithology*; Lindsay, *Science in the Subarctic*; Nute, "Kennicott in the North"; Peattie, "Kennicott"; "Robert Kennicott" (memorial and obituary); Schlachtmeyer, *A Death Decoded*; Sherwood, *Exploration of Alaska 1865–1900*; Thomas, "The Smithsonian and the Hudson's Bay Company"; Turner, "A Century of Biology at Northwestern University"; Vasile, "The Early Career of Robert Kennicott"; Zochert, "Notes on a Young Naturalist." See also Dall and Bannister, "List of the Birds of Alaska," for comments on Kennicott prior to the list of birds (comments at pp. 267–70), and Baird, "On Additions to the Bird-Fauna of North America." Baird and Dall and Bannister all paid homage in their *Transactions of the Chicago Academy of Sciences* publications to Kennicott in the descriptions of what were initially named Kennicott's owl, now the western screech owl (*Megascops kennicotti*), Baird pp. 311–12, and Kennicott's warbler, now variously known as Kennicott's willow warbler, arctic

warbler, Eversmann's warbler, or Alaska willow warbler (see *Phylloscopus borealis* and *Seicercus borealis kennicotti*), Baird pp. 313–15. A subspecies of sharp-tailed grouse found in Canada south of the Great Slave Lake area (*Tympanuchus phasianellus kennicotti*) had previously been named for Kennicott. Of the warbler, Baird said "our bird will, I trust, ever keep the name of Kennicott in recollection" (p. 326).

7. For a popular treatment of the Bartrams, Rafinesque, Audubon, and other early North American naturalists, see Kastner, *A Species of Eternity*. According to curators at the Grove, during his early career as an itinerant portrait painter, Audubon painted portraits of some Kennicott family members then living in New Orleans.

8. Kennicott wrote Roderick McFarlane of the Hudson's Bay Company about life as a Smithsonian curator: "I live constantly here at the Smithsonian among a set of naturalists nearly all of whom have spent themselves in the wilderness" (Kennicott to McFarlane, April 29, 1863, Box 1, Folder 1A, Robert Kennicott Papers, 1863–1865, Smithsonian Institution Archives, Washington, D.C.; hereafter, Kennicott Papers).

9. Kennicott to Hoy, May 7, 1855, cited in Vasile, "The Early Career of Robert Kennicott," p. 162n48. Emphasis in original.

10. Fitzhugh, "Origins of Museum Anthropology at the Smithsonian Institution," p. 180; Deiss, "Spencer F. Baird and His Collectors."

11. Laubacher, "The 'Mandarin-Missionary' Strategy"; Lindsay, ed., *The Modern Beginnings of Subarctic Ornithology*; Lindsay, *Science in the Subarctic*; Sherwood, *Exploration of Alaska 1865–1900*.

12. Fitzhugh, "Origins of Museum Anthropology at the Smithsonian Institution," p. 181.

13. Publications about the 1865–67 Western Union Telegraph Expedition include Adams, *Life on the Yukon 1865–1867*; Adams and Taggart, "Journal of George Russell Adams"; Ahvenainen, *The Far Eastern Telegraphs*, pp. 25–32; Ault, "The (Almost) Russian Telegraph"; Branson, *The Life and Times of John W. Clark*, chapter 2; Deppermann, "Two Cents an Acre"; Galbraith, "Perry McDonough Collins"; James, *The First Scientific Exploration of Russian America*; Kennan, *Tent Life in Siberia*; Landauer, *Scammon beyond the Lagoon*, chapters 6 and 7; Mackay, "The Collins Overland Telegraph"; Neering, *Continental Dash*; Postinov and Falk, *Exploring and Mapping Alaska*, particularly chapter 5; Reid, *The Telegraph in America*, pp. 508–17; Robb, "The Collin's Overland or Russian Extension Telegraph Project"; Taggart and Ennis, "Journal of William H. Ennis"; Vivier, "The Collins Overland Line"; Wilcockson, "Perry Collins's Electronic Rim around the Pacific"; Wympher, *Travel and Adventure*.

14. Postinov and Falk, *Exploring and Mapping Alaska*, pp. 408–12.

15. Vivier, "The Collins Overland Line," p. 243.

16. Latham speech of February 17, 1862, cited in Reid, *The Telegraph in America*, p. 508.

17. Account of testimony found in "Statement accompanying the Memorial of the WUTC for remission of duties on cable purchased for crossing Behring's Straits in the line to Russia now abandoned," no date but likely 1869–70, p. 6, Box 846, Folder

1, Western Union Telegraph Company Records, National Museum of American History, Archives Center, Smithsonian Museum of American History, Washington, D.C. (hereafter, Western Union Records).

18. Clay letters from May 1863, cited in Robb, "The Collin's Overland or Russian Extension Telegraph Project," p. 13n17; see also Postinov and Falk, *Exploring and Mapping Alaska,* p. 414.

19. Ahvenainen, *The Far Eastern Telegraphs,* pp. 158–206; Müller-Pohl, "Wiring the Pacific."

20. Seward to Zachariah Chandler, May 14, 1864, "Communication of Hon. William H. Seward upon the Subject of an Intercontinental Telegraph Connecting the Eastern and Western Hemispheres by way of Behring's Strait." Emphasis in original.

21. Unless otherwise indicated, this paragraph is based on Kennicott to McFarlane, April 15, 1864, Box 1, Folder 1A, Kennicott Papers. Various capitalization, punctuation, and spellings are herein unaltered from the original letter.

22. At the time Kennicott wrote McFarlane, no agreement between the Hudson's Bay Company and Western Union had been reached. In January 1865, the British Colonial Office and the Legislature of British Columbia both approved of the project. In November 1865, Western Union and Hudson's Bay Company reached an agreement for "the services of Hudson's Bay personnel in the construction of the line," presumably in British Columbia (Galbraith, "Perry McDonough Collins," pp. 213–14).

23. No date, but likely early 1864, "supposed Best Route," written notes, Box 1, Folder 24, Hiram Sibley Family Papers, University of Rochester, Rush Rees Library Rare Books, Special Collections, and Preservation Department (hereafter, Sibley Family papers).

24. John Horner to Hiram Sibley, December 25, 1864, Box 1, Folder 4, Sibley Family Papers.

25. "An Act to Encourage and Facilitate Telegraphic Communication between the Eastern and Western Continents," July 1, 1864, 38th Congress, 1st Session. See also certification, August 31, 1864, Box 1, Folder 23, Sibley Family Papers.

26. Stoeckl to Gorchakov, September 6 O.S. (18 N.S.), 1864, Box 1, Folder 22, Sibley Family Papers. Original in French; my grateful appreciation to Sophie Lohmeier for her translation to English.

27. "Statement of the Directors of the Western Union Telegraph Company for the Stockholders," October 1, 1865, Box 1, Folder 26, Sibley Family Papers. For a short summary, see Thompson, *Wiring a Continent,* p. 411.

28. Thomas Eckert to Hiram Sibley (on USMT letterhead), July 19, 1864, Box 1, Folder 3, Sibley Family Papers.

29. "Agreement Salmon P. Chase Secretary of the Treasury with Hiram Sibley," November 1, 1864, Box 1, Folder 21, Sibley Family Papers. Although Western Union was providing a smattering of free telegrams in the greater scheme of things, this telegram patronage explains something I routinely encountered in my research with the diaries, journals, and other documents related to Robert Kennicott, William Dall,

and the scientific expedition. As soon as the scientific expedition was organized in January 1865 and began to travel from Chicago to Washington, on to New York City, and again upon arrival in San Francisco, diaries and other documents report that Kennicott and others were in and out of telegraph offices all day long. Free service from Western Union was fully exercised and also meant telegraph offices were places that members of the scientific expedition often met or went to look for each other. Free use of the Transcontinental Telegraph by the Coast Survey, the Smithsonian, and the Naval Observatory also had been granted by Congress in the 1860 legislation authorizing the Transcontinental Telegraph; see Thompson, *Wiring a Continent,* p. 355.

30. Memo, no date, Smithsonian date stamp March 19, 1865, Box 1, Folder Charles S. Bulkley, Western Union Telegraph Expedition Collection, Smithsonian Institution Archives, Washington, D.C. (hereafter, WUTE Papers).

31. Kennicott to McFarlane, April 15, 1864, Box 1, Folder 1A, Kennicott Papers. Emphasis in original.

32. Clark and Forbes, "Science in Chicago," p. 560; Hiram L. Kennicott, "Historical Sketch of the Academy," in "The Chicago Academy of Sciences Centennial Meeting" pamphlet, May 22, 1957, Box 1, Folder 2, Robert Kennicott and Henry M. Bannister Papers, Northwestern University Archives, Evanston, IL (hereafter, Kennicott and Bannister Papers). The University of Michigan also was involved in this plan for collaborative specimen collection.

33. Robert Kennicott to Northwestern University, no date, 1857, outlining agreement of cooperation between Northwestern University and the Smithsonian Institution and providing instructions on the collection, preservation, and shipping of specimens to Northwestern University, Box 1, Folder 1, Kennicott and Bannister Papers; see also Turner, "A Century of Biology at Northwestern University," p. 5.

34. Dall to Baird, April 4, 1865, Box 1, Folder William H. Dall, WUTE Papers. The involvement of Robert Kennicott in subarctic and arctic fieldwork from 1859 to 1862, of Kennicott and Henry Bannister in the Western Union Telegraph Expedition, and of Oliver Marcy in geological research for a road between the mouth of the Columbia River and Montana Territory together suggest that the encounters between Northwestern University and indigenous peoples in the early decades of Northwestern extended beyond the embroilment of Northwestern University founder, board of trustees president, and Colorado Territory governor John Evans in the Sand Creek Massacre. Encounters also extended beyond the displacement of indigenous peoples locally present when the founders of Northwestern University first acquired land in what is now Evanston. For a detailed study of the relationship between Northwestern and Sand Creek, see "Report of the John Evans Study Committee."

35. Evans to Douglas, 1858, Evanston History Center, Evanston IL, online at http://evanstonhistorycenter.org/past-curators-picks/.

36. Kennicott to Baird, July 23, 1865, Box 1, Folder Robert Kennicott, WUTE Papers. Elision (D—n) in original.

37. Dall to Baird, September 26, 1866, Box 1, Folder William H. Dall, WUTE Papers.

38. "The Collins Overland Telegraph," *Harper's Weekly* August 12, 1865; clipping in Box 31, Folder Collins Overland Route (folder 2 of 7), Western Union Records; Reid, *The Telegraph in America,* pp. 142–46; Thompson, *Wiring a Continent,* pp. 429–39.

39. Myer, "Report Concerning Visit to Cleveland and Conference There with Col. Stager Chief of Mil. Telegs. and Supt. Western Union Telegraph Co.," April 16, 1864, Reel 2, Albert James Myer Papers, Manuscript Division, Library of Congress, Washington, D.C. (hereafter, Myer Papers). Myer intimated in his report that Anson Stager was using the USMT as a vehicle for excessive billing charges to the Union government for the establishment and maintenance of Western Union duplicate and auxiliary telegraph lines.

40. See Plum, *The Military Telegraph,* 2:47–48, 97–100, 208–10.

41. Bulkley to Baird, October 14, 1864, Box 1, Folder Charles S. Bulkley, WUTE Papers.

42. Entry of April 13, 1866, Diary of F. M. Smith, Charles Hubbell Collection, Diaries of F. M. Smith and G. R. Adams, Alaska State Library and Archives, Juneau, AK.

43. Hiram Sibley to Elizabeth Tinker Sibley, October 18 O.S. (30 N.S.), 1864, Box 11, Folder 20, Sibley (Hiram) Family Papers Addition, University of Rochester Rush Rees Library Rare Books, Special Collections, and Preservation Department, Rochester, NY (hereafter, Sibley Family Papers Addition). Hiram Sibley was at the time in St. Petersburg, where he spent several months in 1864 and 1865 negotiating details on the telegraph and cable plans. On Sibley and St. Petersburg, see Coe, *The Telegraph,* pp. 86–87.

44. Telegraph tape, Bulkley to Baird, December 16, 1864, in Box 6, Folder 30, William H. Dall Papers, Smithsonian Institution Archives, Washington, D.C. (hereafter, Dall Papers).

45. "Statement of the Origin, Organization, and Progress of the Russian-American Telegraph Western Union Extension, Collins' Overland Line, via Behring Strait and Asiatic Russia to Europe," May 1866, pp. 11–13, Box 846, Western Union Records.

46. "Puck's Deal," p. 22. "Russias" in original.

47. *New York Examiner and Observer,* "The Girdle of the World."

48. Joseph Henry's Daily Journal, May 23, 1865, Smithsonian Institution Archives, Washington, D.C. (hereafter, Henry Journal); see also May 25, 1865. For a biography of Charles Melville Scammon, see Landauer, *Scammon beyond the Lagoon*; chapters 6 and 7 for the Western Union Telegraph Expedition.

49. Kennicott to Baird, June 4, 1865, Box 1, Folder Robert Kennicott, WUTE Papers. Emphases and spellings in original. The Fraser River is the longest river in British Columbia and generally runs north and south.

50. Kennicott to Baird, July 9, 1865, Box 1, Folder Robert Kennicott, WUTE Papers. Emphasis and spellings in original.

51. Not everyone arrived in Sitka on the same day, because passage to Alaska had taken place on several ships, and not everyone immediately headed for Sitka. Most

who went from San Francisco to Sitka at the start of the expedition arrived at Sitka during the last two weeks of August.

52. William Dall contract of employment with Western Union Telegraph Company, July 1, 1865, Box 18, Folder 15, Dall Papers.

53. Plum, *The Military Telegraph,* 2:346–49.

54. Bulkley to Executive Committee, Western Union, July 25, 1865, volume 2, Journal and Letters of Colonel Charles S. Bulkley, Alaska State Library and Archives, Juneau, AK.

55. William Dall to Charles Dall, September 19, 1865, Box 2, Folder 1, Dall Papers.

56. Details on Bulkley's plans and orders including camp guards, regulations on trading with indigenous peoples, and liquor prohibitions are found in Thompson, *Wiring a Continent,* p. 430. Of Bulkley and many other individuals hired for the expedition, James Reid noted "many of these gentlemen had been in the Army of the United States" and as a result, "had the baptism of fire" (Reid, *The Telegraph in America,* p. 512).

57. Kennicott to Baird, July 23, 1865, Box 1, Folder Robert Kennicott, WUTE Papers.

58. George F. Emmons journal excerpt, entry for October 18, 1867, Alaska State Library and Archives, Juneau, AK. On the transfer ceremony, see Black, *Russians in Alaska 1732–1867,* pp. 285–87.

59. Kennicott to Baird, July 9, 1865, Box 1, Folder Robert Kennicott, WUTE Papers. One surmises Baird, Dall, Kennicott, and others were familiar with the work of Charles Dickens, particularly *Martin Chuzzlewit.*

60. William Dall to Charles Dall, September 19, 1865, Box 2, Folder 1, Dall Papers. Dall and Kennicott were far from alone in this view of Hyde; Joseph Dyer recorded in his diary that Hyde was "dispised [*sic*] by everyone." Entry of August 22, 1865, Joseph Dyer Diaries, Atwood Research Center, Anchorage Museum, Anchorage, AK (hereafter, Dyer Diaries). In addition to silver scallop shells and gold bugs, other military uniform insignia of the Western Union Telegraph Expedition included canoes, snowshoes, and a "W.U." inscription for Western Union Telegraph Company.

61. Log book entry, July 13, 1865, Box 2, Folder 2, Dall Papers. There is a question of why Kennicott chose Dall instead of Bannister, because Bannister had spent more time at the Smithsonian than had Dall. As discussed later in the chapter, Bulkley split the scientific team, and Bannister was not with Kennicott and Dall when Kennicott made his decision.

62. Kennicott to Healey Dall, July 11, 1865, Box 2, Folder 1, Dall Papers.

63. Bannister passage (from his journal) in James, *The First Scientific Exploration of Russian America,* pp. 162–63.

64. Dall to Baird, August 16, 1865, Box 1, Folder William H. Dall, WUTE Papers.

65. William H. Dall Diary, 1865, entries for March 22 and March 25, Smithsonian Institution Archives, Washington, D.C. (hereafter, Dall Diary 1865).

66. Bulkley to Kennicott, September 8, 1865, cited in Postinov and Falk, *Exploring and Mapping Alaska,* p. 419.

67. Bannister to Bannister (father), July 26, 1865, Box 2, Folder 1, Kennicott and Bannister Papers.

68. In addition to the materials cited above, on the splitting of the scientific team by Bulkley see also Schlachtmeyer, *A Death Decoded.*

69. Bannister was born July 24, 1844, and passed away May 1, 1920. Dall was born August 21, 1845, and passed away March 27, 1927.

70. Dall to Baird, August 16, 1865, Box 1, Folder William H. Dall, WUTE Papers.

71. Dall Diary 1865, entry for February 15 (see also March 7 for another visit).

72. "Interview with Dr. Henry M. Bannister," typescript, May 23, 1909, Box 1, Folder 2, Kennicott and Bannister Papers.

73. Dall Diary 1865, "Notes for the Winter 65–66."

74. Woodring, "William Healey Dall, 1845–1927."

75. Now the Garrett-Evangelical Theological Seminary, several of the founders of Northwestern University also founded the Garrett Biblical Institute. The university and institute in the nineteenth century shared some faculty and facilities as well as the emergent Evanston campus. The seminary is now largely independent of the university (save some infrastructure sharing) and remains located on the Evanston campus.

76. James, *The First Scientific Exploration of Russian America,* p. 266.

77. *Northwestern University Bulletin* 16, no. 41 (June 30, 1916), pp. 31–32.

78. Entry of December 13, 1865, Dyer Diaries.

79. Kennicott to Bannister, January 7, 1866, Box 2, Folder 1, Kennicott and Bannister Papers; also in James, *The First Scientific Exploration of Russian America,* p. 274.

80. Ennis to Bannister, January 7, 1866, Box 2, Folder 1, Kennicott and Bannister Papers. Emphasis in original.

81. Loucks, *James Alton James.*

82. "The Ordeal of Edward Spencer," *Chicago Tribune,* September 4, 1960; "Wreck of the *Lady Elgin* and Rescue Work of E. W. Spencer," *Evanston News,* July 6, 1910. Future Northwestern president Henry Fowler also was among Northwestern and Garrett students who helped rescue *Lady Elgin* victims.

83. Willard, *A Classic Town,* pp. 176–79.

84. Bannister, "Funeral Discourse."

85. Henry O. Bannister (father) to Alexander Winchell, May 7, 1862, Box 1, Folder 5, Kennicott and Bannister Papers.

86. Kennicott to (father) Bannister, January 17, 1864, Box 1, Folder 4, Kennicott and Bannister Papers.

87. Kennicott to (son) Bannister, January 18, 1864, Box 1, Folder 4, Kennicott and Bannister Papers. Emphasis in original.

88. See letters from Henry to his father, January 3, 1865, and January 9, 1865, both Box 2, Folder 1, Kennicott and Bannister Papers.

89. Marcy to Bannister, February 20, 1865, Box 2, Folder 1, Kennicott and Bannister Papers. Oliver Marcy later served as president of Northwestern University from 1876 to 1881 and again in 1890.

90. Dall to Charles Dall (father), August 19, 1863, Reel 3, Box 3, Folder 14, Caroline Wells Healey Dall Papers, Massachusetts Historical Society, Boston, MA.

91. Thanks to Mimi White for this brilliant observation, and for her conversations with me about this book.

92. On the shipment of specimens, see Henry Journal, November 23, 1865.

93. Dall Diary, 1866–67, entry for September 25, 1866, Smithsonian Institution Archives, Washington, D.C. (hereafter, Dall Diary, 1866–67).

94. Dall Diary, 1866–67, entry for September 30, 1866.

95. In retrospect, with Bannister departing and some other members of the scientific team soon choosing to either return to San Francisco or to work in British Columbia, this was the moment when "in reality, Dall became the scientific leader of an expedition that lacked researchers" (Postinov and Falk, *Exploring and Mapping Alaska,* p. 428).

96. Kennicott to Baird, June 4, 1865, Box 1, Folder Robert Kennicott, WUTE Papers. The Atlantic cable began accepting public business on August 26, 1866; see Thompson, *Wiring a Continent,* p. 434. By the first two weeks of September 1866, it was clear that the 1866 Atlantic cable was functioning reliably, and an unsuccessful 1865 Atlantic cable had been recovered from the sea floor, repaired and activated, and also placed in service. Suddenly in September 1866 not one, but two, Atlantic cables were in regular operation. See Coe, *The Telegraph,* pp. 100–101.

97. Dall Diary, July–November 1867, entry for July 24, 1867, Smithsonian Institution Archives, Washington, D.C. (hereafter, Dall Diary, July–November 1867). Abbreviations in original. On September 19, 1866, six days before he learned of the death of Kennicott, Dall noted in his diary the recent arrival of some "material and letters" from Victoria, including "news that the Atlantic Cable has been working well for a month." These materials probably arrived on Dall's boat by way of an exchange of letters, newspapers, and other reading material with another passing ship (Dall Diary, 1866–1867). Dall wrote his uncle William in August 1867, "There has been too much stealing somewhere, and the Atlantic Cable dished it" (August 4, 1867, Box 2, Folder 2, Dall Papers).

98. Entry for April 22–23, 1866, Dyer Diaries. Diaries and logbooks of several individuals on the telegraph expedition mention additional instances of hoisting the American flag in Russian Alaska.

99. Thompson, *Wiring a Continent,* pp. 438–39.

100. For summaries of the activities of Bannister and other members of the scientific corps after their Alaska experiences, see Schlachtmeyer, *A Death Decoded,* pp. 134–35.

101. James, *The First Scientific Exploration of Russian America,* pp. 35–45. Expedition member Ferdinand Bischoff, an entomologist, also was in Washington.

102. For Bannister's own recollections, see "Interview with Dr. Henry M. Bannister, May 23, 1909," typescript, Box 1, Folder 2, Kennicott and Bannister Papers.

103. Postinov and Falk, *Exploring and Mapping Alaska*, pp. 433–38.

104. Woodring, "William Healey Dall, 1845–1927."

105. Letter to Commanding General, Department of Alaska, July 13, 1868, RG 393, Letters Received, 1868–70, Box 1, National Archives and Records Administration.

106. *1876 Annual Report of the Chief Signal Officer*, pp. 72–78.

107. On Turner, see ibid., pp. 74–78. On Nelson, see Goldman, "Edward William Nelson." For Nelson's letter (the supply ship was serving the ill-fated *Jeanette*), see Nelson to Bannister, August 21, 1879, Box 2, Folder 1, Kennicott and Bannister Papers.

108. *Memoranda on the Work Accomplished by the Signal Service*, p. 3.

109. "A Chronological Outline of the History of Meteorology in the United States."

110. Monmonier, "Telegraphy, Iconography, and the Weather Map," p. 17.

111. *1879 Annual Report of the Signal Corps*, p. 189.

112. Coe, *The Telegraph*, p. 63; Jessup, "Connecting Alaska." For a contemporary overview, see Hudson, *Connecting Alaskans*.

113. "Farewell Order of the Chief Signal Officer," *1892 Annual Report of the Chief Signal Officer*, p. 341.

114. *1877 Annual Report of the Chief Signal Officer*, p. 31.

115. Smith, "The Most Unforgettable Character I Have Ever Met," no date, typescript, Box 1, Folder 5, Kennicott and Bannister Papers.

116. "Interview with Dr. Henry M. Bannister," typescript, May 23, 1909, Box 1, Folder 2, Kennicott and Bannister Papers.

117. Müller, *Wiring the World*, p. 112. On naturalists and empire in the North Pacific, see Jones, *Empire of Extinction*.

118. Foster, "Robert Kennicott." Parenthetical passage in original.

119. Schlachtmeyer, *A Death Decoded*, gives an excellent account of the forensic examination. See also Bruwelheide et al., "Unearthing Robert Kennicott," pp. 92–123; *Washington Post*, "This Smithsonian Scientist's Death Was a Mystery."

120. Dall collected these remains on April 23, 1867, noting that day he went "down to Massacre Point and open an Indian grave and get a good skull, which I take back and feel quite relieved when it is safely deposited under my bed. As my life may have paid the penalty of discovery" (entry for April 23, 1867, Dall Diary, 1866–1867, p. 101). See also William Healey Dall, Logbook of Collections Sent to the Smithsonian Institution, Western Union Telegraph Expedition, 1868, p. 23, entry "1822. Man from bed of dry lake. L.b. Youkon. R. Am. W. H. Dall"; and see "Inventory of Human Remains" and "Table of Completed Repatriations."

Chapter 5. Hot Winds on a Sun-Baked Desert

1. A Zuni attribute of a rainbow. Zuni weather proverbs were collected in the 1870s and 1880s by Frank Hamilton Cushing, an assistant ethnologist at the Smithsonian. Cushing's research was included (pp. 124–27) in the Signal Service collection "Weather Proverbs."

2. Cushing was with the newly formed Smithsonian Bureau of Ethnology. For collections of writings by Cushing, see Green, ed., *Zuni*; Green, ed., *Cushing at Zuni*. For an assessment of Cushing and his interaction with the Smithsonian, see Parezo, "Cushing as Part of the Team." For an obituary of Cushing, see McGee et al., "In Memoriam: Frank Hamilton Cushing." See also Brandes, "Frank Hamilton Cushing." Henry Farny (see front cover for Henry Farny, *Song of the Talking Wire*) illustrated Cushing's "My Adventures in Zuni."

3. See Scheips, "Albert James Myer."

4. Myer to James Walden, November 15, 1854, entries of December 6, 1854, and January 3, 1855, Reel 1, Albert James Myer Papers, Manuscript Division, Library of Congress, Washington, D.C.

5. Rue, "Pesh-Bi-Yalti Speaks." See also Rue, "Words by Iron Wire."

6. *1879 Annual Report of the Chief Signal Officer*, pp. 132–34, 219–20.

7. *1872 Annual Report of the Chief Signal Officer*, p. 556.

8. *1873 Annual Report of the Chief Signal Officer*, p. 287.

9. Rue, "Words by Iron Wire," p. 6.

10. Rue, "Words by Iron Wire," pp. 6–15. On George Crook, see Greene, "George Crook."

11. *1877 Annual Report of the Chief Signal* Officer, p. 145; Rue, "Words by Iron Wire," pp. 72–75.

12. Wooster, *The Military and United States Indian Policy, 1865–1903*, p. 31.

13. West, *The Last Indian War*, p. 177.

14. Rue, "Words by Iron Wire," pp. 31–32. Rue's passage is based in part from newspaper editorials of March 1872 in the *Weekly Arizona Miner* published in Prescott.

15. Ellis, "Lieutenant A. W. Greely's Report," p. 70.

16. "Protection of Texas Frontier," William Belknap to House of Representatives, April 21, 1874, 43rd Congress, 1st session, H. R. Ex. Doc. 222.

17. Wooster, *Frontier Crossroads*, p. 90.

18. For data on the increase in commercial traffic over military telegraph lines in Arizona Territory, see *1877 Annual Report of the Chief Signal Officer*, pp. 97–103. Further detail on the construction of military telegraph lines in the Southwest during the 1870s and 1880s is found in Ropp, *Hello—Anybody There?*, pp. 26–33; on the uses of the military telegraph by settlers, ranchers, and business interests in the Southwest, see also Tate, *The Frontier Army*, pp. 68–70.

19. Rue, "Words by Iron Wire," pp. 80–81.

20. Ball, *In the Days of Victorio*, p. 37. James Kaywalka confirmed that on occasion Apaches cut the telegraph line of the Mescalero Reservation "for the copper wire which made bright bracelets—not for the purpose of stopping communication. The Apaches needed jewelry for the ceremonies of the young maidens." On the telegraph and Apache surveillance and conflict, see also Collins, *Apache Nightmare*, pp. 25–27. Collins also discusses several instances throughout his book where telegraph lines

were down, sabotaged, or out of commission and on pp. 150–56 discusses detailed telegraphic exchanges between commanding officers at periods when military telegraph lines were in good working order.

21. Rue, "Words by Iron Wire," p. 34.

22. *1872 Annual Report of the Chief Signal Officer*, pp. 523–24.

23. *1875 Annual Report of the Chief Signal Officer*, p. 35. Emphases in original.

24. Ibid., p. 90.

25. *1876 Annual Report of the Chief Signal Officer*, p. 36. Now a ghost town, Indianola was an important Texas port between Galveston and Corpus Christi until it was destroyed by a hurricane in August 1886.

26. *1876 Annual Report of the Chief Signal Officer*, pp. 85–86.

27. *1877 Annual Report of the Chief Signal Officer,* p. 36.

28. *1879 Annual Report of the Chief Signal Officer*, p. 184.

29. Rue, "Words by Iron Wire," pp. 41–45.

30. *1881 Annual Report of the Chief Signal Officer,* pp. 119–20. Settlers and private citizens could and did use military telegraph lines in western North America for nominal rates. The Signal Corps preferred that commercial telegraph traffic travel over commercial telegraph lines whenever such lines were available, and in their absence, offered military telegraph commercial services as a substitute for settlers and others.

31. *1888 Annual Report of the Chief Signal Officer,* p. 20.

32. Leckie and Leckie, *Unlikely Warriors,* pp. 250–52. A microfilm collection from the National Archives and Records Administration, M1381 Headquarters Records of the District of the Pecos, 1878–1881 (hereafter, M1381), contains documents relevant to these activities, including records and logs of telegraph reports.

33. Dillon, "The Arrival of the Telegraph in Texas."

34. *1881 Annual Report of the Chief Signal Officer,* p. 111.

35. Ibid., p. 112.

36. Kathke, "Power Lines."

37. For one example of relocating a military telegraph line, see Grierson to Tingle, January 24, 1879, Reel 1, M1381, National Archives and Records Administration (hereafter, NARA).

38. Ropp, *Hello—Anybody There?*, p. 26, explains that in Arizona and New Mexico Territories "the lines were continually changing as to location and the stations operating on the line, so what was the physical status of a line one year would be almost unrecognizable a few years later."

39. Conner, "History of Weather Observations."

40. 1877 *Annual Report of the Chief Signal Officer,* p. 143.

41. Grierson to Blunt, February 4, 1879, Reel 1, M1381, NARA.

42. Benjamin Grierson discussed the lack of clarity in reciprocal agreements for pursuing Victorio across the Mexican border in his year-end report for 1880. See

Grierson report, December 31, 1880, Reel 1, M1381, NARA. The problematics of hot pursuit of criminals and indigenous peoples back and forth across the Mexican-American border preceded the activities of Victorio; see, for example, "Mexican Border Troubles," Message of the President [Rutherford Hayes] to Congress, November 13, 1877, 45th Congress, 1st session, H.R. Ex. Doc. 13.

43. Dinges, "The Victorio Campaign of 1880"; Leckie and Leckie, *Unlikely Warriors,* pp. 258–68. See also Ball, *In the Days of Victorio.*

44. Wooster, *Frontier Crossroads,* pp. 101–109. On Benjamin Grierson, see Dinges, "Benjamin Grierson," in *Soldiers West*, pp. 295–318; Dinges, "The Victorio Campaign of 1880: Cooperation and Conflict on the United States-Mexico Border," pp. 81–94. On Apaches sabotaging telegraph lines with leather and buckskin strung to look like wires, see Ball, *In the Days of Victorio*, pp. 176–177.

45. *1880 Annual Report of the Chief Signal Officer*, pp. 111–28. Chemjuevis in original.

46. *1873 Annual Report of the Chief Signal Officer*, p. 333.

47. *1879 Annual Report of the Chief Signal Officer,* pp. 132–34.

48. *1881 Annual Report of the Chief Signal Officer,* pp. 926–83.

49. *Astronomical and Meteorological Observations*, pp. 386–87 (for Virginia City); p. 369 (for Berthoud).

50. Barker, "On the Total Solar Eclipse of July 29th, 1878."

51. *1876 Annual Report of the Chief Signal Officer,* p. 75. Turner's report, pp. 74–77, is one of the most extensive reports ever filed by a Signal Service observer and includes great detail on a huge range of weather conditions, information on the growing seasons of plants and vegetables, descriptions of aurorae, patterns of migratory birds, and migrations of whales. Turner later worked with the Smithsonian and published several works on the natural history and ethnology of Alaska and arctic regions. Turner was replaced at Fort St. Michaels by Edward Nelson. Nelson also was a prolific naturalist, ethnographer, and ornithologist. For a report by Turner of an unidentified flying object seen in Alaska by Turner and others, see *1877 Annual Report of the Chief Signal Officer*, entry of November 17, 1876, p. 86.

52. Westdahl report of July 23, 1867, Box 1, Folder Report of Ferdinand Westdahl, Western Union Telegraph Expedition Collection, Smithsonian Institution Archives, Washington, D.C.

53. Bartky, "The Adoption of Standard Time."

54. Willis and Hooke, "Cleveland Abbe."

55. Bartky, "The Adoption of Standard Time," p. 37.

56. *1872 House Appropriations Signal Service Report*, p. 31.

57. Stachurski, *Longitude by Wire,* pp. 89–91; on the personal equation, see pp. 120, 175. See also various materials and newspaper clippings in the Robert W. Karrow Research Papers, Newberry Library, Chicago, IL.

58. Stachurski, *Longitude by Wire,* p. 176.

59. Thomsen, "Timing Space," pp. 123–28.

60. "Report upon the Determination of the Astronomical Co-ordinates of the Primary Stations," pp. 11–12, 80–81.

61. For a summary of the conference, see Blaise, *Time Lord*, pp. 193–213.

62. Kathke, "Power Lines," p. 339.

63. Orton to Myer, October 15, 1872, in Wells, *The Relation of the Government to the Telegraph*, pp. 75–93.

64. *1886 Annual Report of the Chief Signal Officer*, p. 6.

65. Ball, *In the Days of Victorio*, p. 13.

66. For reports on the tests, see "Inclosures to Report on Concerted Heliograph Practice, Department of Arizona, May 1–15, 1890," in RG 111, Entry 32, Box 1, Correspondence Concerning Instruments and Methods of Signaling, 1887–93, NARA.

67. Ropp, *Hello—Anybody There?*, p. 63.

68. Hawes, "The Signal Corps."

69. Guston, "Congressmen and Scientists."

70. Abbe quoted in Hawes, "The Signal Corps," p. 75.

71. Hazen obituary, *Cincinnati Enquirer*, January 17, 1887, Box 8, William Babcock Hazen Papers, National Museum of American History, Archives Center, Smithsonian Museum of American History, Washington, D.C.

72. *1890 Annual Report of the Chief Signal Officer*, pp. 5–6.

73. *1888 Annual Report of the Chief Signal Officer*, p. 6. Homing pigeons were an early manifestation of airborne communication, presaging the emergence of wireless telegraphy and radio communication. There had been experiments with manned tethered balloons, mainly as observation posts, dating back to the Civil War. The homing pigeon also was a harbinger of eventual airborne communication by airplane.

Conclusion

1. Harold Bolce, "The Telephone at the White House," pp. 161–62.

2. "Behind the Scenes at the White House."

3. "One of the Old Guard," p. 225.

4. For an example of the direct involvement of McKinley via telegraph in redirecting troop movements, see transcript of telegraph conversation between White House Telegraph Room and telegraph staff of General Nelson Miles, June 8, 1898, in Box 68, George B. Cortelyou Papers, Manuscript Division, Library of Congress, Washington, D.C. (hereafter, Cortelyou Papers). See also O'Toole, *The Spanish War—An American Epic*, and Britton, *Cables, Crises, and the Press*, especially chapter 11, for discussions of waging war with cables and telegraphs.

5. "Report of Capt. Benjamin F. Montgomery, Signal Corps, in Charge of the Telegraph and Cipher Bureau of the White House," September 27, 1901, excerpt from *Annual Reports of the War Department for the Fiscal Year Ended June 30, 1901*, appendix 11.

6. Greely to McKinley, May 18, 1898, Box 68, Cortelyou Papers.

7. Associated Press resolution of self-censorship, May 17, 1898, Box 68, Cortelyou Papers.

8. Fawcett, "Telegraph and Cipher Bureau of the White House." The graphophone was used on occasion to record important telephone calls for McKinley to listen to, and also telegraph signals (the sound of the key and relay) for transcription.

9. *Daily Mail*, "Ghosts of a Forgotten War," February 8, 2014.

10. By 1900, telegram traffic at the White House Telegraph Room exceeded fifty thousand messages annually ("White House Wires").

11. "One of the Old Guard," p. 225. Montgomery was in the White House Telegraph Room and received the first message from Buffalo when McKinley was assassinated on September 6, 1901; see Hood, ed., *Memories of the White House*, p. 263. Some have argued that Montgomery claimed Ida McKinley (spouse of the president) successfully persuaded her husband to retain the Philippines as an American possession in the aftermath of the Spanish-American War. Louis Gould found the evidence for this event inconclusive, although agreed that Ida "almost certainly encouraged her husband's ultimate inclination toward annexation and expansion" (Gould, *American First Ladies*, p. 192). Ida may have first met Montgomery during the Rutherford Hayes Administration, as Ida and William were close to the Hayes family and visited the White House during the Hayes presidency ("First Lady Biography: Ida McKinley").

12. "Like a Vast Checker-Board," p. 1. The Boxer Rebellion of 1900 and the allied armies at "Pekin" (Beijing) were overseen in the field by future Chief Signal Officer George Scriven, who implemented White House telegraph and cable connectivity with the events in China.

13. Squier, "Submarine Cables in War." See also Squier, "An American Pacific Cable." Squier told the audience at the "American Pacific Cable" presentation that the need for an American Pacific cable ultimately was not colonial but geostrategic, noting that "the Philippine question should not overshadow the larger question—the Eastern question—in the consideration of this project" (p. 668).

14. Kennelly, "Biographical Memoir of George Owen Squier."

15. Pritchett, "Some Recollections of President McKinley."

16. Order No. 13, War Department, Signal Office, September 13, 1898, Box 3, Folder Military Records 1896–1898, George Sabin Gibbs Papers, Manuscript Division, Library of Congress, Washington, D.C.

17. Müller, *Wiring the World*, p. 104.

18. Bolce, "The Telephone at the White House."

19. Cortelyou to Loeb, April 1, 1902, Box 37, Cortelyou Papers.

20. *San Francisco Call*, January 10 and June 7, 1905; *Washington Star*, January 9, 1905, newspaper clipping in Box 35, Cortelyou Papers.

21. Loomis, "The White House Telephone and Crisis Management."

22. For an overview of the Situation Room that places its origins in the Kennedy administration, see Bohn, *Nerve Center*.

23. Fawcett, "The War Room."

24. Lavine, *Circuits of Victory*, p. 13.

25. Ibid., p. 65. The Signal Corps had seen significant reductions in its ranks in the twenty-five or so years before World War I, in part because of events such as the Signal Service transformation into the National Weather Service (now NOAA) and the transition from former military telegraph and telephone networks into commercial service providers.

26. See National Museum of the American Indian, "Native Words, Native Warriors."

27. Kruger, "Tribal Broadband," p. 1.

28. See, for example, Benson, *Wired Wilderness*.

29. See TeleGeography submarine cable map at https://www.telegeography.com/telecom-resources/submarine-cable-map/index.html. For an analysis of undersea cables, see Starosielski, *The Undersea Network*. On global electronic networks, sensors, and environmental measurements, see Gabrys, *Program Earth*.

30. Cortada, *All the Facts*, pp. 148–52.

31. Native Americans were first "identified" in the 1860 census. Despite being identified from 1860 on, they were not enumerated until 1890. That is, census data about Native American populations was collected prior to 1890 but was not incorporated into national population census statistics regardless of taxation status until 1890. A 1921 fire at the Department of Commerce destroyed a significant amount of 1890 census data (Collins, "Native Americans in the Census").

32. U.S. Census Office, 1892, p. xlvii, cited in Jobe, "Native Americans and the U.S. Census."

33. Turner, "The Significance of the Frontier in American History," at https://www.historians.org/about-aha-and-membership/aha-history-and-archives/archives/the-significance-of-the-frontier-in-american-history.

34. Gore, *Earth in the Balance*.

35. Unrau, ed., *Tending the Talking Wire*, p. 235.

Bibliography

Archival Collections, Museums, and Site Visits

All manuscript collections cited herein from the National Archives and Records Administration (NARA) were located (as of April 2015) at the National Archives building at 700 Pennsylvania Avenue, Washington, DC. The diaries and logbooks of William Dall and of Joseph Henry's Daily Journal cited herein are available online in digital transcription and as reproduced original pages at the Smithsonian Digital Volunteers Transcription Center.

Alaska State Library, Archives, and Museum, Juneau, Alaska

American West Photographs Collection, National Archives and Records Administration (NARA), Washington, DC

Atwood Research Center, Anchorage Museum, Anchorage, Alaska

Baranof Castle State Historic Site (Castle Hill, site of 1867 transfer of Alaska), Sitka, Alaska

Bulkley, Charles, Journal and Letters of Colonel Charles S. Bulkley, Alaska State Library, Archives, and Museum, Juneau, Alaska

Cortelyou, George B., Papers, Manuscript Division, Library of Congress, Washington, DC

Dall, Caroline Wells Healey, Papers, Massachusetts Historical Society, Boston, Massachusetts

Dall, William H., Papers, circa 1839–1858, 1862–1927, Smithsonian Institution Archives, Washington DC, including the following diaries and logbooks available online at the Smithsonian Digital Volunteers Transcription Center: Dall, William H., Diary, 1865, Smithsonian Institution Archives, Washington, DC; Dall, William H., Diary, 1866–67, Smithsonian Institution Archives, Washington, DC;

Dall, William H., Diary, July–November 1867, Smithsonian Institution Archives, Washington, DC; Dall, William H., Logbook of Collections Sent to the Smithsonian Institution, Western Union Telegraph Expedition, 1868, Smithsonian Institution Archives, Washington, DC

Dall-Healey Family Photographs, Massachusetts Historical Society Photo Archives, Boston, Massachusetts

Denver Public Library Digital Collections, Denver, Colorado

Dyer, Joseph, Diaries, Atwood Research Center, Anchorage Museum, Anchorage, Alaska

Emmons, George. George F. Emmons Journal Excerpt. Alaska State Library, Archives, and Museum, Juneau, Alaska

Evans, John. Letter to Steven Douglas, 1858. Evanston History Center, Evanston, Illinois

Gibbs, George Sabin, Papers, Manuscript Division, Library of Congress, Washington, DC

Greely, Adolphus Washington (A. W.) Papers, Manuscript Division, Library of Congress, Washington, DC

The Grove (Kennicott family home and National Historic Site), Glenview, Illinois

Hazen, William Babcock, Papers, National Museum of American History, Archives Center, Smithsonian Museum of American History, Washington, DC

Henry, Joseph. Daily Journal, 1865. Smithsonian Institution Archives, Washington, DC

History Colorado Digital Collections, Stephen H. Hart Library and Research Center, Denver, Colorado

Kansas Historical Society, Kansas Memory Digital Collections, Topeka, Kansas

Karrow, Robert W., Research Papers, Newberry Library, Chicago, Illinois

Kennicott, Robert, Papers, 1863–1865, Smithsonian Institution Archives, Washington, DC

Kennicott, Robert, and Henry M. Bannister, Papers, Northwestern University Archives, Evanston, Illinois

Library of Congress Prints and Photographs Online Catalog, Washington, DC

Microfilm collection M1381, Headquarters Records of the District of the Pecos, 1878–1881, National Archives and Records Administration (NARA), Washington, DC

Myer, Albert James, Papers, Manuscript Division, Library of Congress, Washington, DC

National Oceanic and Atmospheric Administration, National Weather Service Collection, Washington, DC

Pope, Frank. J., Reports of Western Union Telegraph Company explorations in western Canada, under Maj. Frank J. Pope 1865–1866, NAA MS 1682, National Anthropological Archives, Smithsonian Institution, Washington, DC

Records of the Department and District of Alaska, 1867–70. Record Group (RG) 393, National Archives and Records Administration (NARA), Washington, DC

Records of the Signal Corps. Record Group (RG) 111, National Archives and Records Administration (NARA), Washington, DC

Records of the United States Senate. Territorial Papers Series, Senate Territorial Papers Relating to Minnesota. Record Group (RG) 46, National Archives and Records Administration (NARA), Washington, DC

Sandbo, Preston O. "Westward to Europe." Undated manuscript (based on bibliography, after 1953). Alaska State Library, Archives, and Museum, Juneau, Alaska

Seward, William Henry, Papers, University of Rochester Rush Rees Library Rare Books, Special Collections, and Preservation Department, Rochester, New York

Sheldon Jackson Museum, Sitka, Alaska

Sibley, Hiram, Family Papers. University of Rochester Rush Rees Library Rare Books, Special Collections, and Preservation Department, Rochester, New York

Sibley (Hiram) Family Papers Addition, University of Rochester Rush Rees Library Rare Books, Special Collections, and Preservation Department, Rochester, New York

Sitka National Historic Park and Russian Bishop's House, Sitka, Alaska

Smith, F. M., Diary, Charles Hubbell Collection, Diaries of F. M. Smith and G. R. Adams. Alaska State Library, Archives, and Museum, Juneau, Alaska

University of Alaska–Fairbanks Digital Archives, Fairbanks, Alaska

Western Union Telegraph Company Records, National Museum of American History, Archives Center, Smithsonian Museum of American History, Washington, DC

Western Union Telegraph Expedition (WUTE) Collection, 1865–1867, Smithsonian Institution Archives, Washington, DC

William Henry Jackson Collection, Scotts Bluff National Monument, National Park Service, Scotts Bluff County, Nebraska

Works Cited

The Annual Reports of the Chief Signal Officer are included within the Annual Reports of the Secretary of War. The Chief Signal Officer reports cited herein are abbreviated as Annual Reports of the Chief Signal Officer. The ongoing digitization of government documents has now seen most, if not all, of these reports digitized and retrievable through various search engines. For some of the reports from the 1880s, I used print copies at the National Archives. The copies I consulted in 2015, however, were pulled for me from a queue of documents to be digitized, so they are now likely available as digitized documents. With the exception of documents found in archival collections (see notes), the government documents, newspaper accounts, and similar items herein were retrieved through online search engines or otherwise found as digitized documents. All websites cited in this book were active as of December 1, 2016.

"An Act to Encourage and Facilitate Telegraphic Communication between the Eastern and Western Continents." July 1, 1864, 38th Congress, 1st Session.

Adams, George R. *Life on the Yukon 1865–1867*. Edited by Richard A. Pierce. Kingston, ON: Limestone, 1982.

Adams, George Russell, and Harold F. Taggart. "Journal of George Russell Adams: Member, Exploring Expedition in Russian America, 1865–67." *California Historical Society Quarterly* 35, no. 4 (December 1956): 291–307.

Ahvenainen, Jorma. *The Far Eastern Telegraphs*. Helsinki, Finland: Suomalainen Tiedeakatemia, 1981.

Allin, Lawrence C. "'A Mile Wide and an Inch Deep': Attempts to Navigate the Platte River." *Nebraska History* 63 (1982): 1–15.

Annual Report of the Chief Signal Officer. Years 1864–1900.

Appropriation for Telegraph and Telephone Service, District of Columbia, March 25, 1897. H.R. Document 11, 55th Congress, 1st session.

Arrington, Leonard. "The Deseret Telegraph." *Journal of Economic History* 11 (1951): 117–39.

Astronomical and Meteorological Observations Made during the Year 1876 at the United States Naval Observatory. Washington, DC: Government Printing Office, 1880.

Ault, Phillip H. "The (Almost) Russian Telegraph." *American Heritage Magazine* 26, no. 4 (June 1975).

Baird, Spencer F. "On Additions to the Bird-Fauna of North America, Made by the Scientific Corps of the Russo-American Telegraph Expedition." *Transactions of the Chicago Academy of Sciences* 1, no. 2 (1869): 311–26.

Ball, Eve. *In the Days of Victorio: Recollections of a Warm Springs Apache*. Tucson: University of Arizona Press, 1970.

Bannister, Henry O. "Funeral Discourse on the Death of Lieutenant H. G. Meacham." Chicago: Dunlop and Sewell, 1863.

Barker, George. "On the Total Solar Eclipse of July 29th, 1878." *Proceedings of the American Philosophical Society* 18, no. 102 (July–December 1878): 103–14.

Bartky, Ian R. "The Adoption of Standard Time." *Technology and Culture* 30, no. 1 (January 1989): 25–66.

Bates, Alice. "The History of the Telegraph in California." *Annual Publication of the Historical Society of Southern California* 9 (1914): 181–87.

Bates, David Homer. *Lincoln in the Telegraph Office*. New York: Century, 1907.

"Behind the Scenes at the White House: The White House Telegraph." *Deseret News*, November 30, 1897, 9.

Bektas, Yakup. "The Sultan's Messenger: Cultural Constructions of Ottoman Telegraphy, 1847–1880." *Technology and Culture* 41, no. 4 (October 2000): 669–96.

Benson, Etienne. *Wired Wilderness: Technologies of Tracking and the Making of Modern Wildlife*. Baltimore: Johns Hopkins University Press, 2010.

Berwanger, Eugene M. *The Rise of the Centennial State: Colorado Territory, 1861–76*. Urbana: University of Illinois Press, 2007.

Billings, Elden E. "Military Activities in Washington in 1861." *Records of the Columbia Historical Society* 60/62 (1960–1962): 123–33.

Binnema, Ted. *Enlightened Zeal: The Hudson's Bay Company and Scientific Networks, 1670–1870*. Toronto, ON: University of Toronto Press, 2014.

Black, Lydia T. *Russians in Alaska 1732–1867*. Fairbanks: University of Alaska Press, 2004.

Blaise, Clark. *Time Lord: Sir Sandford Fleming and the Creation of Standard Time*. New York: Vintage, 2000.

Blondheim, Menahem. *News over the Wires: The Telegraph and the Flow of Public Information in America, 1844–1897*. Cambridge, MA: Harvard University Press, 1994.

———. "'Public Sentiment Is Everything': The Union's Public Communications Strategy and the Bogus Proclamation of 1864." *Journal of American History* 89, no. 3 (December 2002): 869–99.

Bohn, Michael K. *Nerve Center: Inside the White House Situation Room*. Washington, DC: Brassey's, 2003.

Bolce, Harold. "The Telephone at the White House—The President's Private Branch Exchange." *American Telephone Journal* 7, no. 11 (March 14, 1903): 161–62.

Brandes, Ray. "Frank Hamilton Cushing: Pioneer Americanist." Ph.D. dissertation, University of Arizona, 1965.

Branson, John B. *The Life and Times of John W. Clark of Nushagak, Alaska, 1846–1896*. Washington, DC: U.S. Department of the Interior, National Park Service, Lake Clark National Park and Preserve, 2012.

Britton, John. *Cables, Crises, and the Press: The Geopolitics of the New International Information System in the Americas, 1866–1903*. Albuquerque: University of New Mexico Press, 2013.

Brown, J. Willard. *The Signal Corps, USA, in the War of the Rebellion*. Boston: Signal Corps Association, 1896.

Brown, Richard D. *Knowledge Is Power: The Diffusion of Information in Early America, 1700–1865*. New York: Oxford University Press, 1989.

Bruce, Robert V. *1877: Year of Violence*. Indianapolis: Bobbs-Merrill, 1959.

Bruwelheide, Karin S., Sandra S. Schlachtmeyer, Douglas W. Owsley, Vicki E. Simon, Arthur C. Aufderheide, Larry W. Cartmell, and Stephan J. Swanson. "Unearthing Robert Kennicott: Naturalist, Explorer, Smithsonian Scientist." In *Studies in Forensic Biology: Anthropological Perspectives*, edited by Christopher M. Stojanowski and William N. Duncan, 67–91. Cambridge, U.K.: Cambridge University Press, 2017.

Carter, Denny. *Henry Farny*. New York: Watson-Guptill, 1978.

Chicago Tribune. "The Ordeal of Edward Spencer," September 4, 1960.

"A Chronological Outline of the History of Meteorology in the United States of North America." *Monthly Weather Review*, August 1909, 146–49.

Clark, Austin H., and Leila G. Forbes. "Science in Chicago." *Scientific Monthly* 36, no. 6 (June 1933): 556–67.

Coates, Kenneth. "The Kennicott Network: Robert Kennicott and the Far Northwest." *Yukon Historical and Museums Association Proceedings* 2 (1984): 25–32.

Coe, Lewis. *The Telegraph: A History of Morse's Invention and Its Predecessors in the United States*. Jefferson, NC: McFarland, 1993.

Collins, Charles. *Apache Nightmare: The Battle at Cibecue Creek*. Norman: University of Oklahoma Press, 1999.

Collins, James P. "Native Americans in the Census, 1860–1890." *Prologue* 38, no. 2 (Summer 2006), https://www.archives.gov/publications/prologue/2006/summer/indian-census.html.

"Communication of Hon. William H. Seward upon the Subject of an Intercontinental Telegraph Connecting the Eastern and Western Hemispheres by Way of Behring's Strait." Washington, DC: Government Printing Office, 1864.

Conner, Glenn. "History of Weather Observations in Wickenburg, Arizona, 1875–1948," May 2006, Station History Report to the Climate Database Modernization Program of NOAA's National Climatic Data Center, Midwestern Regional Climate Center. [Available from Midwestern Regional Climate Center, 2204 S. Griffith Drive, Champaign, IL 61820.]

Convery, William. "John M. Chivington." In *Soldiers West: Biographies from the Military Frontier*, 2nd ed., edited by Paul Andrew Hutton and Durwood Ball, 149–73. Norman: University of Oklahoma Press, 2009.

Cooling, Benjamin. "Defending Washington during the Civil War." *Records of the Columbia Historical Society* 71–72 (1971–72): 314–37.

Coppedge, Bryan R., and James H. Shaw. "Effects of Horning and Rubbing Behavior by Bison (*Bison bison*) on Woody Vegetation in a Tallgrass Prairie Landscape." *American Midland Naturalist* 138, no. 1 (July 1997): 189–96.

Cortada, James W. *All the Facts: A History of Information in the United States since 1870*. New York: Oxford University Press, 2016.

Coutant, C. G. *The History of Wyoming from the Earliest Known Discoveries*. Laramie, WY: Chaplin, Spafford, and Mathison, 1899.

Cox, William V. "The Defenses of Washington: General Early's Advance on the Capital and the Battle of Fort Stevens, July 11 and 12, 1864." *Records of the Columbia Historical Society* 4 (1901): 135–65.

Cronon, William. *Nature's Metropolis: Chicago and the Great West*. New York: Norton, 1991.

———. "The Trouble with Wilderness; or, Getting Back to the Wrong Nature." In *Uncommon Ground: Rethinking the Human Place in Nature*, edited by William Cronon, 69–90. New York: Norton, 1995.

Curtiss, Daniel S. *Western Portraiture and Emigrant's Guide: A Description of Wisconsin, Illinois, and Iowa*. New York: Colton, 1852.

Cushing, Frank Hamilton. "My Adventures in Zuni." *Century Magazine* (February 1883), 500–511.

Daily Mail. "Ghosts of a Forgotten War: Naval Archivists Discover Trove of Never Before Seen Photographs from Spanish-American Conflict of 1898," February 8, 2014.

Dall, William H., and Henry M. Bannister. "List of the Birds of Alaska, with Biographical Notes." *Transactions of the Chicago Academy of Sciences* 1, no. 2 (1869): 267–310.

Daughters of the American Revolution. *DAR Handbook and National Bylaws*, 31st ed. Washington, DC: National Society Daughters of the American Revolution, 2014.

Deignan, Herbert G. "The HBC and the Smithsonian." *The Beaver* 278:4 (1947): 2–7.

Deiss, William. "Spencer F. Baird and His Collectors." *Journal of the Society for the Bibliography of Natural History* 9, no. 4 (1980): 635–45.

Deppermann, W. H. "Two Cents an Acre." *North American Review* 245 (Spring 1938): 126–33.

"Diary of Jake Pennock." *Annals of Wyoming* 23, no. 2 (July 1951): 4–29.

Dillon, Charles H. "The Arrival of the Telegraph in Texas." *Southwestern Historical Quarterly* 64, no. 2 (October 1960): 200–211.

Dillow, Frank. "Connecting Oregon: The Slow Road to Rapid Communications, 1843–2009." *Oregon Historical Quarterly* 111, no. 2 (Summer 2010): 184–219.

Dinges, Bruce J. "The Victorio Campaign of 1880: Cooperation and Conflict on the United States–Mexico Border." *New Mexico Historical Review* (January 1987): 81–94.

Downey, Gregory. *Telegraph Messenger Boys: Labor, Technology, and Geography, 1850–1950*. New York: Routledge, 2002.

"Durango Gun Club Offers Free Gun Sighting-In to Prevent Insulator Damage." KOB-TV, October 9, 2015. [KOB-TV online story no longer found on KOB-TV website, last accessed April 2016.]

Eckel, R. W., and R. Snow. "Coast Guard Landline Communications: 1873 through 1974." *Coast Guard Engineer's Digest*, January-February-March 1975, 62–70.

Ellis, L. Tuffly. "Lieutenant A. W. Greely's Report on the Installation of Military Telegraph Lines in Texas, 1875–1876." *Southwestern Historical Quarterly* 69, no. 1 (July 1965): 66–87.

Ernst, Erik A. "John A. Kennicott of the Grove: Physician, Horticulturalist, and Journalist in 19th Century Illinois." *Journal of the Illinois State Historical Society* 74, no. 2 (Summer 1981): 109–18.

Evanston News. "Wreck of the *Lady Elgin* and Rescue Work of E. W. Spencer," July 6, 1910.

Fawcett, Waldon. "Telegraph and Cipher Bureau of the White House." *Popular Science News*, September 1902, 203–4.

———. "The War Room at the White House." *The World's Work*, 3, no. 4 (March 1904): 1841–43.

"First Lady Biography: Ida McKinley," http://www.firstladies.org/biographies/firstladies.aspx?biography=25.

Fitzhugh, William. "Origins of Museum Anthropology at the Smithsonian Institution and Beyond." In *Anthropology, History, and American Indians: Essays in Honor of William Curtis Sturtevant*, edited by William Merrill and Ives Goddard, 179–200. Washington, DC: Smithsonian Institution Press, 2002.

Fleming, James R. *Meteorology in America, 1800–1870*. Baltimore: Johns Hopkins University Press, 1990.

———. "Storms, Strikes, and Surveillance: The U. S. Army Signal Office, 1861–1891." *Historical Studies in the Physical and Biological Sciences* 30, no. 2 (2000): 315–32.

———. "Telegraphing the Weather: Military Meteorology, Strategy, and 'Homeland Security' on the American Frontier in the 1870s," In *Instrumental in War: Science, Research, and Instruments between Knowledge and the World*, edited by Steven A. Walton, 153–78. Leiden, Neth.: Brill, 2005.

Foster, J. W. "Robert Kennicott." *Western Monthly* 3, no.15 (March 1870): 165–72.

Gabler, Edwin. *The American Telegrapher: A Social History, 1860–1900*. New Brunswick, NJ: Rutgers University Press, 1988.

Gabrys, Jennifer. *Program Earth: Environmental Sensing Technology and the Making of a Computational Planet*. Minneapolis: University of Minnesota Press, 2016.

Galbraith, John S. "Perry McDonough Collins at the Colonial Office." *British Columbia Historical Quarterly* 17, nos. 3–4 (July–October 1953): 207–14.

Gamble, James. "Early Reminiscences of the Telegraph on the Pacific Coast." *Californian* 3 (1881): 321–26.

———. "Wiring a Continent." *Californian* 3 (1881): 556–63.

Gelderman, Carol. *A Free Man of Color and His Hotel*. Washington, DC: Potomac, 2012.

General Order no. 56, War Department, Adjutant General's Office, 1866.

Gilmore, Matthew. "Washington Goes Wireless: The District's 20-Year Struggle to Put Utility Wires Underground." *InTowner*, October 17, 2015, http://intowner.com/2015/10/17/washington-goes-wireless-the-districts-20-year-struggle-to-put-utility-wires-underground/.

Glover Park History website curated by Carlton Fletcher at http://gloverparkhistory.com/

Goldman, Edward. "Edward William Nelson: Naturalist, 1855–1934." *Auk* 52, no. 2 (April 1935): 135–48.

Gore, Albert Arnold Jr. *Earth in the Balance: Ecology and the Human Spirit*. Emmaus, PA: Rodale, 2006.

Gould, Louis. *American First Ladies: Their Lives and Their Legacy*. New York: Garland, 1996.

Grant, Ulysses S. *Personal Memoirs of U. S. Grant*. New York: Charles L. Webster, 1885–86.

Greely, A. W. "The Military-Telegraph Service." In *The Photographic History of the Civil War . . . Soldier Life, Secret Service*, edited by Francis Trevelyan Miller and Robert Sampson Lerner, 341–68. New York: Review of Reviews, 1911.

Green, Jesse, ed. *Zuni: Selected Writings of Frank Hamilton Cushing*. Lincoln: University of Nebraska Press, 1979.

———, ed. *Cushing at Zuni: The Correspondence and Journals of Frank Hamilton Cushing, 1879–1884*. Albuquerque: University of New Mexico Press, 1990.

Greenberg, Joel. *A Feathered River across the Sky: The Passenger Pigeon's Flight to Extinction*. New York: Bloomsbury, 2014.

Greene, Jerome. "George Crook." In *Soldiers West: Biographies from the Military Frontier*, 2nd ed., edited by Paul Andrew Hutton and Durwood Ball, 246–73. Norman: University of Oklahoma Press, 2009.

Grice, Gary. "History of Weather Observations." Station History Report to the Climate Database Modernization Program of NOAA's National Climatic Data Center, Midwestern Regional Climate Center. [Available from Midwestern Regional Climate Center, 2204 S. Griffith Drive, Champaign, IL 61820.]

Guston, David H. "Congressmen and Scientists in the Making of Science Policy: The Allison Commission, 1884–1886." *Minerva* 32, no. 1 (1994): 25–52.

Hamblin, Jacob Darwin. *Arming Mother Nature: The Birth of Catastrophic Environmentalism*. New York: Oxford University Press, 2013.

Hawes, Joseph M. "The Signal Corps and Its Weather Service, 1870–1890." *Military Affairs* 30, no. 2 (Summer 1966): 68–76.

Headquarters of the Department of the Missouri. Circular 13, Ft. Leavenworth, Kansas. April 24, 1883.

Hendrickson, Ed. "Defending Washington: The District of Columbia Militia, 1861." *Washington History* 23 (2011): 36–58.

Hendrickson, Walter B. "Robert Kennicott, an Early Professional Naturalist in Illinois." *Transactions of the Illinois Academy of Science* 63, no. 1 (1970): 104–106.

Henry, Alfred J. "The Passing of Signal Service, Weather Bureau Electric Telegraph and Cable Systems." *Monthly Weather Review* (June 1929): 246–47.

Henson, Pamela M. "'Objects of Curious Research': The History of Science and Technology at the Smithsonian." *Isis* 90 (1999): 249–69.

"History of the Signal Service with a Catalog of Publications, Instruments, and Stations." Washington, DC: Army Signal Office, 1884.

Hochfelder, David. *The Telegraph in America, 1832–1920*. Baltimore: Johns Hopkins University Press, 2012.

Hood, Henry, ed. *Memories of the White House: The Home Life of Our Presidents from Lincoln to Roosevelt*. Boston: Little, Brown, 1911.

Howe, Daniel Walker. *What Hath God Wrought: The Transformation of America, 1815–1848*. Oxford: Oxford University Press, 2007.

Hudson, Heather. *Connecting Alaskans: Telecommunications in Alaska from Telegraph to Broadband*. Fairbanks: University of Alaska Press, 2015.

Indian Affairs: Laws and Treaties, vol. 2. Washington, DC: Government Printing Office, 1904.

"Inventory of Human Remains from the Nulato Area of Alaska in the Collections of the National Museum of Natural History, Smithsonian Institution." 2005. http://anthropology.si.edu/repatriation/reports/regional/alaska/nulato.htm.

Jackson, Sheldon. "Eleventh Annual Report on Introduction of Domestic Reindeer into Alaska, with Maps and Illustrations, 1901." U. S. Senate document 98, 57th Congress, 1st session. Washington, DC: Government Printing Office, 1902.

James, James Alton. *The First Scientific Exploration of Russian America and the Purchase of Alaska*. Evanston, IL: Northwestern University Press, 1942.

Jepsen, Thomas C. "The Telegraph Comes to Colorado: A New Technology and Its Consequences." *Essays and Monographs in Colorado History* 7, no. 3 (1987): 1–26.

Jessup, David Eric. "Connecting Alaska: The Washington–Alaska Military Cable and Telegraph System." *Journal of the Gilded Age and Progressive Era* 6, no. 4 (October 2007): 384–408.

Jobe, Margaret M. "Native Americans and the U.S. Census: A Brief Historical Survey." *University of Colorado Libraries Faculty and Staff Contributions*. Paper 28. Boulder: University of Colorado Libraries.

John, Richard. *Network Nation: Inventing American Telecommunications*. Cambridge, MA: Belknap, 2010.

Johnson, Albert. "Reminiscences of the Honorable Edwin M. Stanton, Secretary of War." *Records of the Columbia Historical Society* 13 (1910): 69–97.

Jones, Ryan Tucker. *Empire of Extinction: Russians and the North Pacific's Strange Beasts of the Sea, 1741–1867*. Oxford, U.K.: Oxford University Press, 2014.

Kastner, Joseph. *A Species of Eternity*. New York: Knopf, 1977.

Kathke, Torsten. "Power Lines: Arizona Elites, the Telegraph, and the Construction of a Regional Identity, 1870–1910." In *Global Communication Electric: Business, News and Politics in the World of Telegraphy*, edited by M. Michaela Hampf and Simone Müller-Pohl, 331–52. Frankfurt, Ger.: Campus, 2014.

Kennan, George. *Tent Life in Siberia, and Adventures among the Koraks and Other Tribes in Kamchatka and Northern Asia*. New York: Putnam and Sons, 1877.

Kennelly, Arthur. "Biographical Memoir of George Owen Squier." Washington, DC: National Academy of Sciences, 1938.

Kennicott, Robert. "Description of a New Snake from Illinois." *Proceedings of the Academy of Natural Sciences of Philadelphia* 8 (1856): 95–96.

Kielbowicz, Richard B. "The Telegraph, Censorship, and Politics at the Outset of the Civil War." *Civil War History* 40, no. 2 (June 1994): 95–118.

Kruger, Lennard G. "Tribal Broadband: Status of Deployment and Federal Funding Programs." *Congressional Research Service Report*. December 20, 2016.

Landauer, Lyndall Baker. *Scammon beyond the Lagoon: A Biography of Charles Melville Scammon*. Pasadena, CA: Flying Cloud, 1986.

Laubacher, Matthew. "The 'Mandarin-Missionary' Strategy: Robert Kennicott, Spencer Fullerton Baird, and Specimen Collection in the Hudson's Bay Territory." *Endeavor* 36, no. 2 (2012): 46–54.

Lavine, A. Lincoln. *Circuits of Victory*. New York: Doubleday, 1921.

Leckie, William H., and Shirley A. Leckie. *Unlikely Warriors: General Benjamin H. Grierson and His Family*. Norman: University of Oklahoma Press, 1984.

Letters about "underground telegraph and telephone wires" from the Secretary of War and the Commissioners of the District of Columbia to Senate. April 12, 1888, 50th Congress, 1st session, ex. doc. 153.

Lindsay, Debra, ed. *The Modern Beginnings of Subarctic Ornithology: Correspondence to the Smithsonian Institution, 1856–1868*, vol. 10, Winnipeg: Manitoba Record Society Publications, 1991.

———. *Science in the Subarctic: Trappers, Traders and the Smithsonian Institution*. Washington, DC: Smithsonian Institution Press, 1993.

Lockwood, John, and Charles Lockwood. *The Siege of Washington: The Untold Story of the Twelve Days That Shook the Union*. New York: Oxford University Press, 2011.

Loomis, Richard T. "The Telephone Comes to Washington: George C. Maynard, 1839–1919." *Washington History* 12, no. 2 (Fall–Winter 2000–2001): 22–40.

———. "The White House Telephone and Crisis Management." *U. S. Naval Institute Proceedings* (December 1969): 63–73.

Loucks, Vernon. *James Alton James: A Short Biography*. Evanston, IL: Northwestern University Press, 1963.

Lukasik, Stephen J. "Why the ARPANET Was Built." *IEEE Annals of the History of Computing* 33, no. 3 (March 2011): 4–21.

Mackay, Corday. "The Collins Overland Telegraph." *British Columbia Historical Quarterly* 10, no. 3 (July 1946): 187–216.

A Manual of Military Telegraphy for the Signal Service. Washington, DC: Government Printing Office, 1872.

McDermott, John D. *Circle of Fire: The Indian War of 1865*. Mechanicsburg, PA: Stackpole, 2003.

McGee, W. J., et al. In Memoriam: Frank Hamilton Cushing." *American Anthropologist*, n.s. 2, no. 2 (April–June 1900): 354–80.

Memoranda on the Work Accomplished by the Signal Service of the United States. Washington, DC: Government Printing Office, 1881.

"Mexican Border Troubles." Message of the President [Rutherford Hayes] to Congress, November 13, 1877. 45th Congress, 1st session, H.R. Ex. Doc. 13.

Meyn, Susan Labry. *Henry Farny Paints the Far West*. Cincinnati, OH: Cincinnati Art Museum, 2007.

Midwestern Regional Climate Center Climate Database Modernization Program. "Histories of Weather Observation at Selected Locations," http://mrcc.sws.uiuc.edu/FORTS/histories1.jsp.

Mihelich, Dennis N., and James E. Potter, eds. *First Telegraph Line across the Continent: Charles Brown's 1861 Diary*. Lincoln: Nebraska State Historical Society.

Millard, Candace. *Destiny of the Republic: A Tale of Madness, Medicine, and the Murder of a President*. New York: Anchor, 2012.

Miller, Eric. "New Light on the Beginnings of the Weather Bureau from the Papers of Increase A. Lapham." *Monthly Weather Review*, February 1931, 65–70.

Monmonier, Mark. *Air Apparent: How Meteorologists Learned to Map, Predict, and Dramatize Weather*. Chicago: University of Chicago Press, 1999.

———. "Telegraphy, Iconography, and the Weather Map: Cartographic Weather Reports by the United States Weather Bureau, 1870–1935." *Imago Mundi* 40 (1988): 15–31.

Müller, Simone M. "From Cabling the Atlantic to Wiring the World." *Technology and Culture* 57, no. 3 (July 2016): 507–26.

———. *Wiring the World: The Social and Cultural Creation of Global Telegraph Networks*. New York: Columbia University Press, 2016.

Müller-Pohl, Simone. "Wiring the Pacific: North American Perspectives on a (De)colonial Project." In *Provincializing the United States: Colonialism, Decolonization, and (Post)Colonial Governance in Transnational Perspective*, edited by Ursula Lehmkuhl, Eva Bischoff, and Norbert Finzsch, 155–80. Heidelberg, Ger.: Universitätsverlag Winter, 2014.

Natale, Simone. "Understanding Media Archeology." *Canadian Journal of Communication* 37 (2012): 523–27.

"National Archives Celebrates the 145th Anniversary of Nevada Statehood." NARA press release. September 23, 2009, http://www.archives.gov/press/press-releases/2009/nr09-127.html.

National Museum of the American Indian, Smithsonian Institution. "Native Words, Native Warriors" [webpage]. http://www.nmai.si.edu/education/codetalkers/html/.

National Park Service. *Santa Fe Trail Map Timeline 1866–1873*, https://www.nps.gov/safe/learn/historyculture/map-timeline-3.htm.

Neering, Rosemary. *Continental Dash: The Russian-American Telegraph*. Victoria, BC: Horsdal and Schubart, 1989.

New York Examiner and Observer. "The Girdle of the World," November 30, 1865.

New York Times. "Action of the Government," July 26, 1877, 5.

———. "Hayes's Administration," March 2, 1881.

———. "Signal Service Proficiency," August 10, 1877.

Nickles, David. *Under the Wire: How the Telegraph Changed Diplomacy*. Cambridge, MA: Harvard University Press, 2003.

Nielson, P. Raymond. "Edward Creighton and the Pacific Telegraph." *Mid-America* 24, no. 1 (January 1942): 61–74.

Northwestern University Bulletin 16, no. 41 (June 30, 1916): 31–32.

Nute, Grace. "Kennicott in the North." *Beaver* 274 (September 1943): 28–32.

Observation and Report of Storms by Telegraph and Signal for the Benefit of Commerce. Washington, DC: Government Printing Office, 1871.

"One of the Old Guard." *Telegraph Age*, 19 (1902): 225.

O'Toole, George. *The Spanish War—An American Epic*. New York: Norton, 1984.

Parezo, Nancy J. "Cushing as Part of the Team: The Collecting Activities of the Smithsonian Institution." *American Ethnologist* 12, no. 4 (November 1985): 763–74.

Peattie, Donald Culross. "Kennicott: Knight of the North." *Esquire* (March 1936): 94–128.

Peck, Graham. "New Records of the Lincoln-Douglas Debate at the 1854 Illinois State Fair: The *Missouri Republican* and the *Missouri Democrat* Report from Springfield." *Journal of the Abraham Lincoln Association* 30, no. 2 (Summer 2009): 25–80.

Peters, Arthur King. *Seven Trails West*. New York: Abbeville, 1996.

Pfeiffer, David A. "Records Relating to the U.S. Military Railroads during the Civil War." *Prologue* 43, no. 2 (Summer 2011), http://www.archives.gov/publications/prologue/2011/summer/usmrr.html.

Phalen, William J. *How the Telegraph Changed the World*. Jefferson, NC: McFarland, 2014.

Plum, William R. *The Military Telegraph during the Civil War in the United States*. Chicago: Jansen, McClurg, 1882.

Postinov, Alexy, and Marvin Falk. *Exploring and Mapping Alaska: The Russian America Era, 1741–1867*. Translated by Lydia Black. Fairbanks: University of Alaska Press, 2015.

Pound, Roscoe. "The Military Telegraph and the Civil War." *Proceedings of the Massachusetts Historical Society*, 3rd ser., v. 66 (Oct. 1936–May 1941): 185–203.

Pritchett, Henry. "Some Recollections of President McKinley and the Cuban Intervention." *North American Review* 189, no. 640 (March 1909): 397–403.

"Protection of Texas Frontier." William Belknap to House of Representatives, April 21, 1874. 43rd Congress, 1st session, H. R. Ex. Doc. 222.

"Puck's Deal." *The Telegrapher*, November 28, 1864, 22.

Raines, Rebecca Robbins. *Getting the Message Through: A Branch History of the U.S. Signal Corps*. Washington, DC: Government Printing Office, 2011.

Regulations for the Operation and Maintenance of the United States Military Telegraph Lines. Washington, DC: Government Printing Office, 1899.

Reid, James D. *The Telegraph in America: Its Founders, Promoters, and Noted Men*. New York: Derby, 1879.

Reilly, Hugh J. *The Frontier Newspapers and the Coverage of the Plains Indian Wars*. Santa Barbara, CA: Praeger, 2010.

Report from House Committee on Appropriations. "Signal-Service and Telegraph Companies." May 9, 1872, 42nd Congress, 2nd session, Report 69.

Report of the Architect of the Capitol to the Secretary of the Interior about November 6, 1898, gas explosion in the Capitol and subsequent repairs. July 1, 1899.

"Report of the John Evans Study Committee." Office of the Provost, Northwestern University. May 2014. PDF at http://www.northwestern.edu/provost/committees/equity-and-inclusion/study-committee-report.pdf.

Report of the Proceedings in the Case of the United States vs. Charles J. Guiteau, Tried in the Supreme Court of the District of Columbia, Holding a Criminal Term, and Beginning November 14, 1881. Washington, DC: Government Printing Office, 1882.

Report on underground telegraphic cables laid between the Capitol and Executive Departments. 48th Congress, 1st session, ex. doc. 197.

"Report upon the Determination of the Astronomical Co-Ordinates of the Primary Stations at Cheyenne, Colorado Territory, and Colorado Springs, Colorado Territory, made during the Years 1872 and 1873." Washington, DC: Government Printing Office, 1874–81.

Resolution of the Legislative Assembly. Territory of Nebraska. February 13, 1865.

Robb, Stewart Andrew. "The Collin's Overland or Russian Extension Telegraph Project: A Pioneer Attempt to Establish Telegraphic Communications between North America and Europe." M.A. thesis, Simon Fraser University, Burnaby, BC, 1968.

"Robert Kennicott." *Transactions of the Chicago Academy of Sciences* 1, no. 2 (1869): 133–226.

Root, George, and Russell K. Hickman. "Pike's Peak Express Companies, Part 4: The Platte Route—Concluded." *Kansas Historical Quarterly* 14, no. 1 (1946): 36–92.

Ropp, Robert O. *Hello—Anybody There? A Brief History of the Military Telegraph System, 1871–1891*. Fort Collins, CO: Old Army Press, 1990.

Rue, Norman L. "Pesh-Bi-Yalti Speaks: White Man's Talking Wire in Arizona." *Journal of Arizona History* 12, no. 4 (Winter 1971): 229–62.

———. "Words by Iron Wire: Construction of the Military Telegraph in Arizona Territory, 1873–1877." M.A. thesis, Department of History, University of Arizona, 1967.

Saturday Globe (Utica, NY). "Like a Vast Checker-Board," March 23, 1901, 1.

Scharf, J. Thomas. *History of St. Louis City and County*. Philadelphia: Everts, 1883.

Scheips, Paul. "Albert James Myer, an Army Doctor in Texas, 1854–1847." *Southwestern Historical Quarterly* 82 (July 1978): 1–24.

———. "Union Signal Communications: Innovation and Conflict." *Civil War History* 9, no. 4 (December 1963): 399–421.

Schiller, Herbert. *Mass Communications and American Empire*. Boulder, CO: Westview, 1992.

Schlachtmeyer, Sandra Spatz. *A Death Decoded: Robert Kennicott and the Alaska Telegraph, a Forensic Investigation*. Belleville, NJ: Voyage, 2010.

Schwartzlose, Richard. *The Nation's Newsbrokers*. Evanston, IL: Northwestern University Press, 1989.

Schwoch, James. *The American Radio Industry and Its Latin American Activities, 1900–1939*. Urbana: University of Illinois Press, 1990.

Sherwood, Morgan B. *Exploration of Alaska 1865–1900*. New Haven, CT: Yale University Press, 1965.

———. "Science in Russian America, 1741–1865." *Pacific Northwest Quarterly* 58, no. 1 (January 1967): 33–39.

"Signal-Service and Telegraph Companies." May 9, 1872. 42nd Congress, 2nd session, Report 69, p. 31.

Singh, Simon. *The Code Book: The Evolution of Secrecy from Mary, Queen of Scots, to Quantum Cryptography*. New York: Doubleday, 1999.

Smith, E. D. Jr. "Communication Pioneering in Oregon." *Oregon Historical Quarterly* 39, no. 4 (December 1938): 352–71.

Soll, Jacob. *Information Master: Jean-Baptiste Colbert's Secret State Intelligence System*. Ann Arbor: University of Michigan Press, 2009.

Spencer, J. A. *History of the United States from the Earliest Period to the Administration of Andrew Johnson*, vol. 4. New York: Johnson, Fry, 1866.

Squier, George. "An American Pacific Cable." Paper presented at the 138th meeting of the American Institute of Electrical Engineers, New York, New York, December 27, 1899.

———. "Submarine Cables in War." In *Abstract of the Course for 1900*, ed. Naval War College, 49–53. Washington, DC: Government Printing Office, 1901.

Stachurski, Richard. *Longitude by Wire: Finding North America*. Columbia: University of South Carolina Press, 2009.

Standage, Tom. *The Victorian Internet: The Remarkable Story of the Telegraph and Nineteenth Century's On-Line Pioneers*. New York: Walker, 1998.

Starosielski, Nicole. *The Undersea Network*. Durham, NC: Duke University Press, 2015.

Starr, Paul. *The Creation of the Media: Political Origins of Modern Communication*. New York: Basic Books, 2004.

Sunder, John E. "Arkansas' First 'Wonder Working Wire.'" *Arkansas Historical Quarterly* 16, no. 3 (Autumn 1957): 231–42.

———. "The Early Telegraph in Rural Missouri, 1847–1859." *Missouri Historical Review*, October 1956, 42–53.

———. "St. Louis and the Early Telegraph, 1847–1857." *Missouri Historical Review*, April 1956, 248–58.

———. "Telegraph Beginnings in Kansas." *Kansas Historical Quarterly* (1959): 32–38.

"Table of Completed Repatriations." October 31, 2014. Smithsonian Institution, National Museum of Natural History, Repatriation Office. http://anthropology.si.edu/repatriation/pdf/table_of_repatriations_2014.pdf.

Taggart, Harold F., and William H. Ennis. "Journal of William H. Ennis: Member, Russian-American Telegraph Exploring Expedition." *California Historical Society Quarterly* 33, no. 1 (March 1954): 1–12.

Tate, Michael. *The Frontier Army in the Settlement of the West*. Norman: University of Oklahoma Press, 1999.

TeleGeography submarine cable map, https://www.telegeography.com/telecom-resources/submarine-cable-map/index.html.

"The Telegraphic Historical Society of North America." Washington, DC: W. F. Roberts. 1894 report.

"Telegraph-Offices in the Corridors of the Capitol." House of Representatives, 43rd Congress, 1st session, misc. doc. 269.

Thomas, Greg. "The Smithsonian and the Hudson's Bay Company." *Prairie Forum* 10, no. 2 (1985): 283–306.

Thompson, Robert Luther. *Wiring a Continent: The History of the Telegraph Industry in the United States, 1832–1866*. Princeton, NJ: Princeton University Press, 1947.

Thomsen, Gene A. "Timing Space: Determining the Western Boundary of Nebraska." *Nebraska History* 80 (1999): 123–28.

"Tragedy of April 14, 1865, as recalled by Eight Survivors of the U. S. Military Telegraph Corps." *Milwaukee Sentinel*, April 11, 1915.

Turner, Clarence. "A Century of Biology at Northwestern University." *Beta Beta Beta Biological Society* 25, no. 1 (March 1954): 2–34.

Turner, Frederick Jackson. "The Significance of the Frontier in American History." Paper delivered at American Historical Association annual meeting, Chicago, 1893; see the version at https://www.historians.org/about-aha-and-membership/aha-history-and-archives/archives/the-significance-of-the-frontier-in-american-history.

Unrau, William E., ed. *Tending the Talking Wire: A Buck Soldier's View of Indian Country, 1863–1866*. Salt Lake City: University of Utah Press, 1979.

Unterman, Katherine. *Uncle Sam's Policemen: The Pursuit of Fugitives across Borders*. Cambridge, MA: Harvard University Press, 2015.

Van Orsdel, Ralph A. "History of the Telephone System in the District of Columbia." *Records of the Columbia Historical Society* 48–49 (1946–47): 171–80.

Vasile, Ronald S. "The Early Career of Robert Kennicott, Illinois' Pioneering Naturalist." *Illinois Historical Journal* 87, no. 3 (Autumn 1994): 150–70.

Vivier, Charles. "The Collins Overland Line and American Continentalism." *Pacific Historical Review* 28 (1959): 247–53.

Wagner, Alexandra M. "Grasshoppered: America's Response to the 1874 Rocky Mountain Locust Invasion." *Nebraska History* 89, no. 4 (2008): 154–67.

Ware, Eugene. *The Indian War of 1864*. New York: St. Martins, 1960.

Washington Post. "This Smithsonian Scientist's Death Was a Mystery: 150 years later, His Skeleton Helped Solve It," August 3, 2016.

"Weather Proverbs." Washington, DC: Government Printing Office, 1883.

Wells, David. *The Relation of the Government to the Telegraph*. New York: 1873.

West, Elliott. *The Essential West: Collected Essays*. Norman: University of Oklahoma Press, 2012.

———. *The Last Indian War: The Nez Perce Story*. New York: Oxford University Press, 2009.

Wheeler, Tom. *Mr. Lincoln's T-Mails: How Abraham Lincoln Used the Telegraph to Win the Civil War*. New York: Harper Collins, 2006.

"White House Wires." *Telegraph Age*, 19, no. 1 (1901): 523–24.

Wilcockson, Stephen. "Perry Collins's Electronic Rim around the Pacific: The Russian-American Telegraph, 1865–1867." M.A. thesis, University of British Columbia–Vancouver, 1996.

Willard, Frances E. *A Classic Town: The Story of Evanston by "An Old Timer."* Chicago: Women's Christian Temperance Union, 1891.

Willis, Edmund P., and William H. Hooke. "Cleveland Abbe and American Meteorology, 1871–1901." *Bulletin of the American Meteorological Society* (March 2006): 315–26.

Wilson, William B. *A Glimpse of the United States Military Telegraph Corps and of Abraham Lincoln*. Harrisburg, PA: Meyers, 1889.

Winkler, Jonathan. *Nexus: Strategic Communications and Security in World War I*. Cambridge, MA: Harvard University Press, 2008.

Wisnek, Dwayne R., and Robert M. Pike. *Communication and Empire: Media, Markets, and Globalization, 1860–1930*. Durham, NC: Duke University Press, 2007.

Wolff, Joshua. *Western Union and the Creation of the American Corporate Order*. New York: Cambridge University Press, 2013.

Woodring, W. P. "William Healey Dall, 1845–1927." Washington, DC: National Academy of Sciences, 1958.

Wooster, Robert. "Fort Davis and the Close of a Military Frontier." *Southwestern Historical Quarterly* 110, no. 2 (October 2006): 172–91.

———. *Frontier Crossroads: Fort Davis and the West*. College Station: Texas A&M University Press, 2006.

———. *The Military and United States Indian Policy, 1865–1903*. New Haven, CT: Yale University Press, 1988.

Wympher, Frederick. *Travel and Adventure in the Territory of Alaska*. New York: Harper and Brothers, 1869.

Wynes, Charles E. "James Wormley of the Wormley Hotel Agreement." *Centennial Review* 19, no. 1 (Winter 1975): 397–401.

Yanoff, Steven. *The Second Mourning: The Untold Story of America's Most Bizarre Political Murder*. Bloomington, IN: AuthorHouse, 2014.

Zochert, Donald. "Notes on a Young Naturalist." *Audubon*, March 1980, 34–47.

Zorrow, William Frank. "Jeptha H. Wade in California: Beginning the Transcontinental Telegraph." *California Historical Society Quarterly* 29, no. 4 (December 1950): 345–56.

Index

Page numbers in *italics* refer to illustrations.

JAMES SCHWOCH is a professor of communication studies at Northwestern University. He is the author of *The American Radio Industry and Its Latin American Activities, 1900–1939* and *Global TV: New Media and the Cold War, 1946–69*.

THE HISTORY OF COMMUNICATION

Selling Free Enterprise: The Business Assault on Labor and Liberalism, 1945–60 *Elizabeth A. Fones-Wolf*

Last Rights: Revisiting *Four Theories of the Press* *Edited by John C. Nerone*

"We Called Each Other Comrade": Charles H. Kerr & Company, Radical Publishers *Allen Ruff*

WCFL, Chicago's Voice of Labor, 1926–78 *Nathan Godfried*

Taking the Risk Out of Democracy: Corporate Propaganda versus Freedom and Liberty *Alex Carey; edited by Andrew Lohrey*

Media, Market, and Democracy in China: Between the Party Line and the Bottom Line *Yuezhi Zhao*

Print Culture in a Diverse America *Edited by James P. Danky and Wayne A. Wiegand*

The Newspaper Indian: Native American Identity in the Press, 1820–90 *John M. Coward*

E. W. Scripps and the Business of Newspapers *Gerald J. Baldasty*

Picturing the Past: Media, History, and Photography *Edited by Bonnie Brennen and Hanno Hardt*

Rich Media, Poor Democracy: Communication Politics in Dubious Times *Robert W. McChesney*

Silencing the Opposition: Antinuclear Movements and the Media in the Cold War *Andrew Rojecki*

Citizen Critics: Literary Public Spheres *Rosa A. Eberly*

Communities of Journalism: A History of American Newspapers and Their Readers *David Paul Nord*

From Yahweh to Yahoo!: The Religious Roots of the Secular Press *Doug Underwood*

The Struggle for Control of Global Communication: The Formative Century *Jill Hills*

Fanatics and Fire-eaters: Newspapers and the Coming of the Civil War *Lorman A. Ratner and Dwight L. Teeter Jr.*

Media Power in Central America *Rick Rockwell and Noreene Janus*

The Consumer Trap: Big Business Marketing in American Life *Michael Dawson*

How Free Can the Press Be? *Randall P. Bezanson*

Cultural Politics and the Mass Media: Alaska Native Voices *Patrick J. Daley and Beverly A. James*

Journalism in the Movies *Matthew C. Ehrlich*

Democracy, Inc.: The Press and Law in the Corporate Rationalization of the Public Sphere *David S. Allen*

Investigated Reporting: Muckrakers, Regulators, and the Struggle over Television Documentary *Chad Raphael*

Women Making News: Gender and the Women's Periodical Press in Britain *Michelle Tusan*

Advertising on Trial: Consumer Activism and Corporate Public Relations in the 1930s *Inger L. Stole*

Speech Rights in America: The First Amendment, Democracy, and the Media *Laura Stein*
Freedom from Advertising: E. W. Scripps's Chicago Experiment *Duane C. S. Stoltzfus*
Waves of Opposition: The Struggle for Democratic Radio, 1933–58 *Elizabeth Fones-Wolf*
Prologue to a Farce: Democracy and Communication in America *Mark Lloyd*
Outside the Box: Corporate Media, Globalization, and the UPS Strike *Deepa Kumar*
The Scripps Newspapers Go to War, 1914–1918 *Dale Zacher*
Telecommunications and Empire *Jill Hills*
Everything Was Better in America: Print Culture in the Great Depression *David Welky*
Normative Theories of the Media *Clifford G. Christians, Theodore L. Glasser, Denis McQuail, Kaarle Nordenstreng, and Robert A. White*
Radio's Hidden Voice: The Origins of Public Broadcasting in the United States *Hugh Richard Slotten*
Muting Israeli Democracy: How Media and Cultural Policy Undermine Free Expression *Amit M. Schejter*
Key Concepts in Critical Cultural Studies *Edited by Linda Steiner and Clifford Christians*
Refiguring Mass Communication: A History *Peter Simonson*
Radio Utopia: Postwar Audio Documentary in the Public Interest *Matthew C. Ehrlich*
Chronicling Trauma: Journalists and Writers on Violence and Loss *Doug Underwood*
Saving the World: A Brief History of Communication for Development and Social Change *Emile G. McAnany*
The Rise and Fall of Early American Magazine Culture *Jared Gardner*
Equal Time: Television and the Civil Rights Movement *Aniko Bodroghkozy*
Advertising at War: Business, Consumers, and Government in the 1940s *Inger L. Stole*
Media Capital: Architecture and Communications in New York City *Aurora Wallace*
Chasing Newsroom Diversity: From Jim Crow to Affirmative Action *Gwyneth Mellinger*
C. Francis Jenkins, Pioneer of Film and Television *Donald G. Godfrey*
Digital Rebellion: The Birth of the Cyber Left *Todd Wolfson*
Heroes and Scoundrels: The Image of the Journalist in Popular Culture *Matthew C. Ehrlich and Joe Saltzman*
The Real Cyber War: The Political Economy of Internet Freedom *Shawn M. Powers and Michael Jablonski*
The Polish Hearst: *Ameryka-Echo* and the Public Role of the Immigrant Press *Anna D. Jaroszyńska-Kirchmann*
Acid Hype: American News Media and the Psychedelic Experience *Stephen Siff*
Making the News Popular: Mobilizing U.S. News Audiences *Anthony M. Nadler*

Indians Illustrated: The Image of Native Americans in the Pictorial Press
John M. Coward

Mister Pulitzer and the Spider: Modern News from Realism to the Digital
Kevin G. Barnhurst

Media Localism: The Policies of Place *Christopher Ali*

Newspaper Wars: Civil Rights and White Resistance in South Carolina, 1935–1965
Sid Bedingfield

Across the Waves: How the United States and France Shaped the International Age of Radio *Derek W. Vaillant*

Race News: Black Reporters and the Fight for Racial Justice in the Twentieth Century *Fred Carroll*

Becoming the Story: War Correspondents since 9/11 *Lindsay Palmer*

Wired into Nature: The Telegraph and the North American Frontier *James Schwoch*

The University of Illinois Press
is a founding member of the
Association of American University Presses.

Composed in 10.5/13 Minion Pro
with Frutiger LT Std display
by Lisa Connery
at the University of Illinois Press
Manufactured by Sheridan Books, Inc.

University of Illinois Press
1325 South Oak Street
Champaign, IL 61820-6903
www.press.uillinois.edu